Analysis and Design of Next-Generation Software Architectures

T0213292

Arthur M. Langer

Analysis and Design of Next-Generation Software Architectures

5G, IoT, Blockchain, and Quantum Computing

 Springer

Arthur M. Langer
Center for Technology Management
Columbia University
New York, NY, USA

ISBN 978-3-030-36901-9   ISBN 978-3-030-36899-9   (eBook)
https://doi.org/10.1007/978-3-030-36899-9

This Springer imprint is published by the registered company Springer Nature Switzerland AG
The registered company address is: Gewerbestrasse 11, 6330 Cham, Switzerland

Foreword

Why Do We Need a Next Generation of Analysis and Design?

The Early Days of Analysis and Design

When the first Analysis and Design methodologies were invented in the 80s (SA/SD in the US, SSADM in the UK and Merise in France), they were intended to support the building of business applications which would last for a long time with minimum cost of maintenance. They were to ensure designs which would cope with the functional (the what) and non-functional (volume and performance) system requirements, which were supposed to stay as stable as possible. Data would be organized to represent required information according to well-defined data models which optimized data access and reduced required storage volumes. The famous relational data model imposed integrity constraints which were maintained on a permanent basis; and transactions were carefully controlled in order to avoid inconsistency in the event of hardware failure or performance bottleneck.

Analysis and Design delivered robust and maintainable applications where consistency and performance were built around invariants identified and defined by the analyst. Analysts expended significant time and effort to build information system that was resistant to change and for which success was measured by the length of their lifecycle, conformance to requirements, and delivery on time and on budget.

Analysts were creating digital objects in the great engineering tradition of building bridges, cars or aircrafts. In the world of engineers, investment in time delivered value is creating long-lasting assets. This mentality was replicated in the Business-to-Business world where business processes were stable and where companies pursued long term objectives that were to be managed through their information systems.

Analysis and Design in the Digital Era

The digital era which started with the advent of internet brought a whole different perspective on time to the development, adoption and evolution of IT applications.

Business to Consumer applications emerged on PC's in the 2000s, and then on smartphones in the 2010s. Suddenly IT was not constrained to bringing progress through a well-defined set of objectives, as it had been previously for business systems. Instead it was offering successive waves of innovation to consumers, provided of course that consumers embraced their adoption.

The world of Digital is a Darwinian one where you need to start small, be adopted by consumers and grow fast, or else be rejected. Perhaps the most significant success stories are epitomized by the rapid success of the GAFA,[1] built around the digital phenomenon of *data platforms*. Platforms embody one of the most significant business model disruptions of the Digital Era.

Why are Platforms Different?

Google, Amazon, Uber, and Airbnb have all adopted platform principles to support their disruptive business models—models which have led to a new type of market that shares common core characteristics with the payment card ecosystem.

With the advent of payment cards, a new market type emerged which has been termed "multisided" by economists. In the specific case of cards, we have two sides: (1) the consumer who is the cardholder and (2) the merchant which is offering payment services via a point of sale terminal capable of reading the card and capturing transaction details.

This economic model was formulated by French economist Jean Tirole, an invention for which he was awarded the Nobel prize for economy.

In a nutshell, a multi-sided market can only really take off once each side reaches a critical size; in the payment card example, no one wants to carry a particular card if it is not widely accepted, and no merchant would be willing to invest in a point of sale terminal if there were not many consumers who would use it.

This chicken and egg stalemate can be broken if a market player is able to offer a platform to both sides of the ecosystem with the commitment to pay very little, if anything, for the use of the platform until critical size is reached. It is the platform that is the tool which enables multisided markets.[2]

In the example of Amazon, the platform is successful for customers because they have a single route to find just about everything they need; it is successful for merchants because they enjoy access to many more potential customers than they would without having their presence on Amazon. In the Airbnb model, the two sides are consumers and hotels; while with Uber, they are consumers and drivers.

It can be difficult to reach critical size on each side of the platform, but once achieved, multisided markets will naturally tend to grow to a monopoly position, driving the need for regulation to maintain competition.

In term of systems analysis and design, we need to distinguish between building the platform and building applications for the platform.

---

[1]Google, Amazon, Facebook, Apple—an acronym used to synonymize the digital giants.
[2]Deliberately Digital—Rewriting Enterprise DNA for Enduring Success (Tardieu et al.) Springer. To be published 2020.

Building the Platform

The unprecedented challenge of launching a platform is that of being able to effectively deliver services when its number of users is small on each side, and their trading volumes are low. The initial cost to users of participating in the platform needs to reflect low startup value, even though this means that the platform operator might have to run at a loss until user revenues build up. But probably even more challenging is the problem of being able to maintain the same architecture for the platform while the two sides are growing from zero to critical size.

For an engineer, it is like building a low cost bridge which only needs to carry a few trucks at the beginning, but which is then progressively reinforced to carry many more trucks as demand rises. Not an easy task!

Building Applications for Platforms

Relatively few companies will take the risk of building a new platform from scratch, and those that do will tend to be large or rapidly growing. However, many companies will seek to join established multisided markets and will try to design and build applications on and around the supporting platforms to offer new value add services.

The very nature of platforms in terms of scalability, performance, security and privacy means that rigid architectural frameworks do not give the necessary freedom to appropriately design new applications. The analysis of the available platform-hosted data combined with the player's own data and target business model brings the real value-add in shaping new services to offer to consumers.

The Importance of Digital Moments

We have traditionally differentiated between business applications which can cope with a latency of a few seconds or more, and technical applications that require real time operation with a latency in the low milliseconds range. To guarantee low latency, technical applications require real time loops that are painstakingly optimized by scrutinizing code to ensure that they deliver low latency response, whatever the circumstances. In business applications, response time for transactions was a focus of attention and was being guaranteed up to a certain number of users and a maximum volume of data.

The digital era is bringing a new context where consumers have become very impatient whatever the circumstances: they are not interested by how many users are connected, how much data needs to be accessed, or their network latency (which is most of the time dictated by the Internet where there is no guaranteed performance). The only "must" with digital applications is to offer the appropriate quality of service, user experience or "Digital moments". These can be very obvious when streaming music or video—constant buffering and interruptions due to latency quickly render the experience unacceptable. Understanding Digital Moments in the context of things like ordering a car or reserving a hotel room requires more sophistication; and in the case of navigation operation in cars, it is very complex. The nature of the Digital moment is directly linked to the attention span of the consumer and the decisions they will be required to make. Failure to respect Digital moments is a fundamental reason for users to reject services on offer, and the matter should therefore be at the top of analysts' and designers' attention.

Distributed Data

Performance of the internet, huge volumes of data, network latency and security have forced analysts and developers to revisit the topic of data localization. The formerly sacrosanct principle of the data base community: *no redundancy*, is now seriously challenged. From simple caching mechanisms used by content distribution networks such as Akamai (which stores copies of data close to where they are the most frequently used), we have progressed to redundancy by design. Instead of hitting the latency constraints of bringing data to processes, we have started to send processes to data (close to the place where it has been collected). We are experiencing edge computing combining with the internet of things to steadily create a world where most of the data will be stored at the edge and not centrally.

With the relaxation of the "no redundancy" principle comes the counter challenge of data consistency across all locations where data copies are stored. Requiring full consistency is very inefficient because it requires that every update in a given location be replicated in every other location before proceeding to the next update.

Finally, for operational reasons, data needs to be backed up in such a way it can be recovered in the event of failure. One piece of data can end up getting replicated three times: for caching reasons when it is only accessed in read-only mode; for latency reason when performance at the edge is required; and finally, for back-up recovery. It is little surprise that the volume of data to be stored is doubling every 18 months. It remains to be seen how such data growth will be managed.

The Consumer Electronics Attitude and the Weekly (if not Daily) scrum

Consumer electronics has fueled the desire for new devices, even before the previous versions are obsolete. Changing smartphones almost every year while updating operating system releases every 2 months have almost become the norm. Such programmed obsolescence is part of the dynamic of the Business to Consumer market.

Ongoing adoption and usage of platform services is essential to their success, driving a compelling need for designers to pay critical attention to ongoing user relevance, value and buy-in. The need for continuous innovation fuels the practice of building platform functionality gradually, at each iteration (called sprints) adding new features and correcting bugs. The most advanced platforms have daily scrums and the question of new feature releases is like taking the bus, if you don't get on this bus, there be another along in the next 5 min. This offers the ability to incrementally build platform functionality and, in case of problems, to easily go back to the previous release.

UX is the Application

UX design has become a key topic in the Digital era. It is intended to enhance user satisfaction with an application by improving its usability, accessibility and desirability. Since adoption and usage of the application is the ultimate yardstick for success and hence survival, a lot of attention should be paid to the user interface.

In the past, the design of the user interface tended to be done after the business process has been defined. It was seen as the way to collect the necessary data and then to present outputs for the user to make decisions.

This sequence has been turned on its head with user interfaces being designed first and then progressively augmented to incorporate business logic. Multiple interface channels including voice, virtual reality and haptics are now primary drivers of the application, with business logic and processing being seen somewhat as the hidden aspect of the application.

Adhering to API is Not Optional

Applications that offer services within existing platforms need to respect their architectural principles, which are often reflected in a set of mandatory APIs (e.g., AWS offers a little more than 20). These API's guarantee that when an application is integrated into the platform, it will not disturb performance, alter security or put multitenancy at risk. Failing to respect API protocols will lead to application rejection by the platform operator.

API's are also critical to gain access to data which has been collected by other sides of the platform. APIs are becoming the lingua franca which (e.g.,) allows Fintech companies to have access to customer data collected by an incumbent bank, or to reuse automotive data collected via connected cars. Security and privacy issues will lead all players participating in a platform ecosystem to accept the rules and regulations set up by independent third parties responsible for enforcing data regulation compliance.

"We Know More Than We Can Tell"

Michael Polanyi the famous Hungarian-British polymath used this statement to express the important truth that no analyst can entirely capture the knowledge about a given process or decision. Beyond what can be formally captured in business logic algorithms, there exist rules and patterns which can only be discovered and incorporated through deep learning. Artificial intelligence is unlikely to ever be as efficient and performant as hand coding, but for simple decision making, it can replace manual programming, and for unexpected and non-forecasted situations, it can be a good substitute for the last exit error message. Especially in sophisticated situations unknown to the analyst, AI can bring reasonable suggestions for returning back to the basic philosophy of the process.

The Next Paradigms

The book will elaborate on the appropriate methodologies to answer the above challenges.

At this stage, let us introduce some of the new paradigms which have become commonplace in the digital world when applied to the business to consumer market and which are gradually penetrating the business application environment.

Webscale Computing encapsulates a hardware architecture where computing performance and storage capacity can be increased without stopping operations. *Multitenancy* is the ability to host multiple tenants and their users on the same platform without any side effects in terms of performance, security and operational continuity.

Webscale Computing and Multitenancy are two foundational pillars for cloud computing.

Data distribution is necessary to achieve performance, but it also raises the challenge of updating multiple instantiations of the same data. The approach taken in the digital era is to dispense with permanent data integrity and to settle for eventual consistency. Platform operators will accept working with inconsistent data but will commit that within a reasonable timeframe, it will be able to reestablish consistency. Eventual consistency may not be adequate for some applications, such as managing the booking of airplane seats, but will be good enough for many other applications.

Agile Development has built the foundations of flexible analysis and design by accepting many changes to an application and implementing them quickly after they have been requested. Agile development requires solid architecture principles to build applications iteratively without compromising performance and maintainability. Applications developed using Agile methodologies need to progress to operations as part of a next scrum. This requires a "Devops" attitude where operation departments are able to accept (e.g.,) weekly changes even though they might disrupt continuity of operations.

Many companies have entered the digital era for analysis and design; it requires a different type of organization which is proving difficult to coexist with the traditional model. The book will detail the what and the how for next generation of Analysis and Design.

Paris, France   Hubert Tardieu
                                                                                                       Atos

Preface

The objective of the *Next Generation of Analysis and Design* is to provide managers and practitioners with a guide to building new architectures in response to the power provided by 5G mobile communications. Indeed, 5G will initiate abilities to communicate in a wireless world, allowing Internet of Things (IoT) to become the "data aggregator" for massive amounts of information that will be collected over distributed networks powered by the internet. Unfortunately, the speed of 5G and the proliferation of IoT creates the need for better security. That is, the ability to collect an abundance of valuable information comes at a price. Simply put, our existing architectures are not capable of providing the necessary security to protect the valuable data that we can use to explore new ways of using artificial intelligence (AI) and machine learning (ML). As a result, new architectures must be developed that use the ledger-based approaches offered by blockchain design. Further, these new architectures will also need a new approach in the ways data is stored across complex mobile networks, which has given birth to advanced capabilities of cloud computing.

In order to compete in a global world that require complex supply chains that span everywhere, companies need to take on the re-architecting of their legacy systems. It seems overwhelming to think that 50 years of developing products and services under a specific development mentality can now somehow be changed in a short amount of time. Nevertheless, my epistemology on this issue is very basic, redo your systems or perish as a business. While this may seem harsh and one-dimensional, I believe the costs for not moving quickly will dramatically affect the ability for any organization to compete in the digital age. This book provides the guidance from both a technical perspective and a management approach. My philosophy, which is shared by others, is that it is impossible to redo all existing legacy systems within most organizations. I, as others, offer a "complimentary approach" that suggests that businesses continue to provide the backend services using their legacy products but need to build new ones that offer the new services that are required in a consumer-driven world—a world which I coin as the "consumerization of technology." We cannot expect organizations to cure 50 years of bad architecture overnight. We experienced the challenges and costs of moving to integrating systems using enterprise resource planning (ERP) products that took 20 years to complete. I believe the new migration to mobility will take even longer and at a much higher cost. So we must transform old systems while building new ones while ensuring the connectivity between them. That is the objective of the *Next Generation of Analysis and Design*.

From a management perspective, executives need to drive a new culture. According to Gupta (2018), successful digital strategies need to emphasize a complimentary business approach as opposed to attempting to launch independent units or labs—it appears that these initiatives have historically failed to provide effective results. Gupta's framework, while not unique, suggests four basic components for effective cultural migration:

1. Reimagine your business
2. Reevaluate your value chain
3. Reconnect with customers
4. Rebuild your organization.

While I agree with all of these steps, executives still need to rebuild their architectures. There is no question that this new digital era provides more user friendly and user intuitive applications, but make no mistake, the technology is more complex and advanced as ever. Thus, we need technically qualified leaders who understand how to build these new systems to support a digital strategy. While cultural transformation is necessary, we must acknowledge that the successful digital-born companies have built backend and front-end systems that operate seamlessly. So, who in the organization can we call our most valuable player (MVP)? My vote in this book is the *Analyst*, the individual who gives you the most payback for the investment. Analysts typically can provide the technical architecture design, usually know their legacy systems, and can perform the necessary project management. All these functions can drive these new systems while helping evolve a new digital-based culture. While we need executives, users, and consumers to be fundamentally engaged in all aspects of the transformation, the analyst represents the one role that can be the main director of its success. Therefore, technology executives need to expand the role of the analyst and understand the importance of the position.

However, this book also recognizes that complimentary approaches only offer short-term solutions. Rather than risk permanent dependencies on old systems, I consistently operate within the auspices of the economic *s-curve*, which so cleverly defines the life-cycle of any product or service. Successful companies need to replace systems by abiding by the curve and start replacing systems before they become obsolete. *The Next generation of Analysis and Design* integrates the S-curve with the SDLC and offers a new way of continually evolving computer architectures. Most important is my prediction that the S-curve will continue to shrink, that is, less time to develop competitive systems, and less time to enjoy their advantages.

It should not be a mystery that corporations today need to be technology centric when developing their competitive strategy. This book also addresses the need to integrate multiple generations of management and staff with a particular focus on the millennial generation (Gen Y). We predict that millennials will be stepping into management positions much quicker than their predecessors and in order to be more competitive in the digital age, companies must do a better job of understanding and assimilating their talents. These assimilations require an integration among baby boomers who are typically the executives and line managers who are often Gen X.

The Next Generation of Analysis and Design also recognizes the role of the consumer. I predict that this era will be known as the "consumer revolution" because of the knowledge consumers have on how digital technologies can provide them with value. These consumer values are often presented by their demand for products and services based on alternative choices and need for personalization. Companies need to recognize that they must offer more product and service choices to multiple and segmented markets to survive in the digital age.

New City, NY, USA   Arthur M. Langer
2020

Acknowledgements

I want to acknowledge my expanding family, DeDe, Michael, Dina, Lauren, Anthony and Lauren P. And of course Cali and Shane Caprio, my two grandchildren who bring me so much joy as they grow.

A special recognition to the alumni of the Executive MS in Technology Executive Masters in Technology Management program for all their great accomplishments that make us all proud.

New City, NY, USA   Arthur M. Langer
January 2020

Contents

Introduction

1

1.1 Traditional Analysis and Design Limitations

Since the beginning of systems development, analysts and designers have essentially adhered to an approach that requires the interview of users, creating logical models, designing them across a network and developing the product. We have gone through for sure different generations, particularly with the coming of client/server systems where we first had to determine what software would reside on the server and what made more sense to stay on the client. A lot of those decisions had to do with systems performance capabilities. Once the Internet became the foundation of application communications and functionality, server technology became the preferred method of designing systems because of version controls and distribution across new devices. Unfortunately, these generations led us to creating a cyber Frankenstein monster. Although security in the mainframe system remains fairly secure, the distributed products across the internet were not designed with enough security. Indeed, the consequences of the lack of security focus has launched the dark web and the crisis of cyber exposures throughout the world. Our Frankenstein monster has created problems beyond our wildest imaginations, affecting not just our systems, but our moral fabric, our laws, our war strategy, and most of all our privacy. As with the Frankenstein novel, the monster cannot be easily destroyed, if at all—such is our challenge. Bottom line, our existing systems, based on central databases and client server mentality cannot protect us, and never will.

Thus, this book is about the next generation of systems architecture, which first must require the unravelling of the monster. This means that all existing systems must be replaced with a new architecture that no longer solely depends on user input, and must be designed to consider what consumers might want in the future and to always consider security exposure. Next Generation Analysis and Design then is a book that takes on this seemingly overwhelming task of the rebuilding of our legacy applications, integration of our new digital technologies, and a security focus that ensures our networks can maximize protection of those that use them.

© Springer Nature Switzerland AG 2020
A. M. Langer, *Analysis and Design of Next-Generation Software Architectures*,
https://doi.org/10.1007/978-3-030-36899-9_1

The good news is that we are on the horizon of getting the new tools and capabilities in which this mission can be accomplished, notwithstanding how long it might take. These capabilities start with the coming of 5G in 2019, which will enable networks to perform at unprecedented speeds. This performance enhancement will drive significant proliferation of Internet of Things (IoT) which in turn will require the creation of massive networks. These networks will need maximum security. To maximize security protection our systems will need to move away from the central database client/server paradigm towards a more ledger-based and object-oriented distributed network that are based on blockchain architecture and cloud interfaces. In order to address the latency exposure of blockchain architecture, some form of quantum computing will be needed. This book will provide an approach or roadmap to accomplishing this transition, particularly the redesign of existing legacy systems.

1.2 Consumerization of Technology in the Digital Age

When the Internet emerged as a game-changing technology, many felt that this era would be known as the Internet Revolution. As digital started to become a common industry cliché, it seemed more certain that "Internet" might be replaced by "Digital." However, upon further analysis, I believe that this revolution will be known historically as the "Consumer" Revolution. The question is why? It appears that the real effects of the internet and the coming of digital technologies has created a population of smart consumers, that is, consumers who understand how technology affords them a more controlling piece of the supply and demand relationship. Indeed, the consumer is in control. The results should be obvious; consumer preferences are changing at an accelerated rate and causing suppliers to continually provide more options and more sophisticated products and services. As a result, businesses must be more agile and "on demand" in order to respond to what Langer (2018) referred to as Responsive Organizational Dynamism (ROD), defined as the measurement of how well organizations respond to change.

This consumerization in the digital era means that analysis and design will need to originate more from a consumer perspective. This means that analysts must expand their requirements gathering beyond the internal user community and seek a greater portion based on consumer needs and more importantly buying habits. Let's examine this point further. The most significant shift in creating new software applications will not only be based on current user needs, but future consumer trends. This represents new cycles of demand in the consumer market. The new demand is based on the close relationship between consumer business needs and home use of digital products and services. From a design perspective, business and home requirements must be blended seamlessly—the ultimate representation of digital life in the 21st century!

But consumerization of technology requires a much more significant leap in design; predictive analytics driven by artificial intelligence (AI) and machine learning (ML) for example. For it is AI and ML that will give us the ability to predict future consumer behavior using a more robust and automated paradigm. Thus, systems must be designed to evolve, just like living forms. In other words, applications must contain what I call *architectural agility*. The first step in architectural agility, is to apply digital re-engineering. The world of application development can only be accomplished by creating enormous object libraries that contain functional primitive operations (Langer 1997). Functional primitive objects are programs that perform the very basic operations, that is, those that provide one simple operation. Basic functional operating programs can be pieced together at execution time to provide more agile applications. By having these primitive objects coming together at execution, it also allows for easier updating of new features and functions. These dynamic linkages at execution provides more evolutionary and agile systems. The object paradigm is not new; the difference in architectural agility is that these objects must be decomposed to its simplest functions. What has previously limited the creation of primitives has been execution latency or performance issues. The performance limitations related to the ability of networks and operating systems to dynamically link primitives to meet performance requirements.

Previous inhibiters to the design of functional primitive objects were incompatibilities between hardware and software environments, which are continually evolving to address this problem. I think we would agree that disfunction among architectures still exist. Just ask people who still experience challenges between Microsoft and Apple systems. Certainly, Steve Jobs can be credited with revolutionizing the consumer interface when he designed a new Apple architecture based on the IPOD and phone designs. This design represents devices that perform less-specific applications but serviced a future wireless-based architecture that could perform more on-demand operations to meet consumer needs. Ultimately Consumerization of Technology treats business applications, personal needs, and everyday life as one integrated set of operations. The Apple architecture then has been at the forefront of an evolutionary platform that can evolve new hardware and software much quicker and efficiently than prior computer systems. In order to keep up with an accelerating evolving consumer it is of utmost importance that businesses focus on how they will transform their legacy applications into this agile digital framework.

1.3 The Role of the Evolving Analyst

Building on the previous section, digital re-engineering represents the challenge of transforming legacy architecture to meet more of the demands of the consumer. As a result, the process of re-engineering, in general, is no longer limited to just working with traditional internal users, rather it must integrate both communities in

any needs assessment. Furthermore, analysis must not only include existing consumer needs, but those that might be the trends of the future! For example, below
are six approaches that were presented in my earlier publication (Langer 2016):

1. *Sales/Marketing*: these individuals sell to the company's buyers. Thus, they
   have a good sense of what customers are looking for, what things they like about
   the business, and what they dislike. The power of the sales and marketing team
   is their ability to drive realistic requirements that directly impact revenue
   opportunities. The limitation of this resource is that it still relies on an internal
   perspective of the consumer.
2. *Third-Party Market Analysis/Reporting*: there are outside resources available
   that examine and report on market trends within various industry sectors. Such
   organizations typically have massive databases of information and using various
   search and analysis can provide various views and behavior patterns of the
   customer base. They can also provide competitive analysis of where the company sits with respect to alternative choices and why buyers may be choosing
   alternative solutions. The shortfall of this approach is that often the data may not
   be specific enough to feed requirements of what applications systems might be
   required to make a competitive advantage for the business.
3. *Predictive Analytics*: this is a hot topic in today's competitive landscape for
   businesses. Predictive analytics is the process of feeding off large datasets and
   predicting future behavior patterns. Predictive analytics approaches are usually
   handled internally with assistance from third-party products or consulting services. The value of predictive analytics is using data to design systems that can
   provide what might be future consumer needs. The limitation is one of risk—the
   risk that the prediction does not occur as planned.
4. *Consumer Support Departments*: Internal teams and external (outsourced
   managed service) vendors that support consumers have a good pulse on their
   preferences because they speak with them. More specifically, they are
   responding to questions, handling problems and getting feedback on what is
   working. These support departments typically depend on applications to help the
   buyer. As a result they are an excellent resource for providing up-to-date things
   that the system does not provide to them that consumers want as a service or
   offering. Often, however, consumer support organizations limit their needs to
   what they experience as opposed to what might be future needs as a result of
   competitive forces.
5. *Surveys*: analysts can design surveys (questionnaires) and send them to consumers for feedback. Using surveys can be of significant value in that the
   questions can target specific application needs. Survey design and administration can be handled by third-party firms, which may have an advantage in that
   the questions are being forwarded by an independent source that might not
   identify the company. On the other hand, this might be considered a negative—
   it all depends on what the analyst is seeking to obtain from the buyer.

6. *Focus Groups*: This approach is similar to the use of a survey. Focus groups are commonly used to understand consumer behavior patterns and preferences. They are often conducted by outside firms. The difference between the focus group and a survey is (1) surveys are very quantitative based using scoring mechanisms to evaluate outcomes. Consumers sometimes may not understand the question and as a result provide distorted information, (2) focus groups are more qualitative and allow analysts to engage with the consumer in two-way dialogue.

Figure 1.1 reflects a graphical depiction of the sources of the analysis of consumers.

Table 1.1 further articulates the methods and deliverables that analysts should consider when developing specifications.

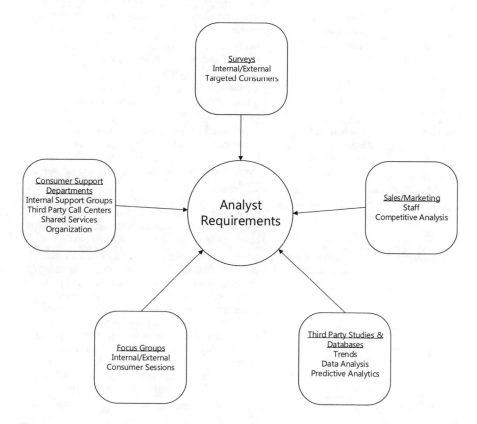

Fig. 1.1 Sources for the analysis of consumers

Table 1.1 Analyst methods and deliverables for assessing consumer needs

Analyst's sources	Methods	Deliverables
Sales/Marketing	Interviews	Should be conducted in a similar way to typical end user interviews. Work closely with senior sales staff. Set up interviews with key business stakeholders
	Win/Loss sales reviews	Review the results of sales efforts. Many firms hold formal win/loss review meetings that may convey important limitations of current applications and system capabilities
Third Party Databases	Document reports reviews	Obtain summaries of the trends in consumer behavior and pinpoint shortfalls that might exist in current applications and systems
	Data analysis	Perform targeted analytics on databases to uncover trends not readily conveyed in available reports
	Predictive analytics	Interrogate data by using analytic formulas that may enable predictive trends in consumer behavior
Support Department	Interviews	Interview key support department personnel (internal and third party) to identify possible application deficiencies
	Data/Reports	Review call logs and recorded calls between consumers and support personnel to expose possible system deficiencies
Surveys	Internal and external questionnaires	Work with internal departments to determine application issues when they support consumers. Use similar surveys with select populations of customers to validate and fine-tune internal survey results
		Use similar surveys targeted to consumers who are not customers and compare results. Differences between existing customer base and non-customers may expose new trends in consumer needs
Focus Groups	Hold internal and external sessions	Internal focus groups can be facilitated by the analyst. Select specific survey results that had unexpected results or mixed feedback and review those results with the focus group attendees. Internal attendees should come from operations management and sales. External focus groups should be facilitated by a third-party vendor and held at independent sites. Discussions with customers should be compared with internal focus group results. Consumer focus groups should also be facilitated by a professional third-party firm

Source Guide to Software Development: Designing and Managing the Life Cycle, 2016

1.4   Developing Requirements for Future Consumer Needs

Perhaps the biggest challenge of the 5G to IoT evolution will be determining what future consumers might want. The question is how to accomplish this challenge? The change brought on by digital inventions will be introduced to incredibly large numbers of consumers in an unprecedented short period of time. Let us just take an historical look at the amount of time it took to reach 50 million consumers.

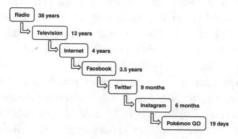

From 38 years to 19 days depicts the incredible acceleration that digital technologies have created. Thus, consumers become aware very quickly and how they respond to new offerings is very much unknown. For example, did Steve Jobs really know that the Mac would primarily be used as a desktop publishing computer when it was designed and first introduced to the consumer market? And did we know that the IPad would be so attractive to executives? The answer of course is no, and remember "almost" is equivalent to "no" in this example. Ultimately analysis and design will evolve to more predictive requirements and will as a result have a failure rate! The concept of risk analysis will be discussed further in Chap. 2. Ultimately, analysis and design has transitioned to being more about collecting data than about self-contained application systems. This transformation is fueling the need for this newly constructed systems architecture.

1.5   The New Paradigm: 5G, IoT, Blockchain, Cloud, Cyber, and Quantum

This section will outline the components of change to the architecture of systems and briefly describe how each component relates to a new and more distributed network of hardware and software components.

1.5.1   5G

While 5G mobile networks and systems clearly are the next generation in global telecommunications it more importantly represents a profound evolution of home, machine-to-machine, and industrial communication capabilities. These new performance capacities will allow for major advancements in the way we utilize artificial intelligence driven by machine learning, and in general how we learn and interact in every part of our lives. Thus, 5G is the initiator of the next generation of systems architecture. This new architecture will be based on enhanced wireless connectivity through distributed networks.

Today approximately 4.8 billion people use mobile services globally. This represents almost two-thirds of the world's population. Connectivity is expected to surpass 5.6 billion people by the end of 2020. Given that many parts of the world have limited physical network infrastructure, enhanced mobile communications represents the only viable approach to linking networks of data and applications. So, 5G is the impetus that fuels the new economy of the future that will be driven by sophisticated mobile communications. This inevitably will create a global economy driven by wireless mobility. Ultimately, 5G is the enabler—an enabler that will allow for specialized networks to participate in what I call "global systems integration" of seamless components. It also represents a scalability of networks that can be dynamically linked and integrated across consumers, communities, corporations, and government entities. This integration will allow these multiple systems to communicate through a common platform to service all aspects of an individual's life. Figure 1.2 provides a graphic depiction of this new reality made possible by 5G performance improvements.

The effects of Fig. 1.2 on analysis and design is significant in that it broadens the scope and complexity of consumer needs and integrates them with all aspects of life. Table 1.2 shows the expansion of coverage to obtain maximum requirements for any product.

Ultimately 5G provides the better performance across wireless networks that requires much more complex systems design. The better performance enables far more complex datasets that can be communicated among multiple types of systems. Most important will be the enablement of mobile devices to utilize these complex datasets across wireless devices. This will in turn drive a whole new economy based on mobility. Mobility will accelerate innovation needs as shown in Fig. 1.3.

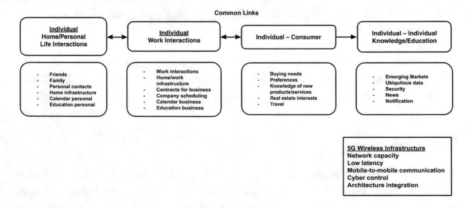

Fig. 1.2 5G mobile connectivity ecosystem enablement

Table 1.2 Scope of analysis and design requirements under 5-G

User/Consumer coverage	A&S response	Comments
Business to Business (BtoB)	Internal user and security	Current process but lacks security process
Business to Consumer (BtoC)	Internal user and external consumer and security	Current but not well integrated in most organizations
Consumer to Consumer (CtoC)	Rare except in specific trading platforms	Needs newer platforms and mobile to mobile
Business to Government	Rare except limited to information, submission of documents, and payments	Overhaul of government and business systems
Individual to Government	Rare except limited to information, submission of documents, and payments	Smart city and compliance driven
Individual to Consumer (ItoC)	Related to member portals	Limited mostly to Facebook/Linkedin
Individual to Individual (ItoI)	Knowledge based portals	Communities of practice and portals of knowledge

Fig. 1.3 Innovation integrated with technology and market needs

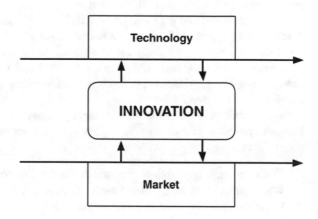

Figure 1.3 shows an interesting cycle of innovations as it relates to the creation of new products and services. The diagram reflects that 5G performance innovations can will create new markets—like mobile applications. On the other hand, the wireless market will in turn demand more features and functions in technology, that need to respond—such as new features and functions that consumers want to see in their apps. So 5G will drive new and more advanced needs across the new frontier of wireless operations.

One challenge of 5G innovation will be the importance it places on moving and linking legacy applications. That is, how will organizations convert their existing systems to compete with more sophisticated "born-digital" products? Furthermore,

5G increased performance allows application developers to better integrate multiple types of datasets, including the proliferation of pictures, videos, and streaming audio services. We expect the existence of non-text-based data to increase by 45% from 2015 to 2020 which will result in a forecasted growth in mobile traffic from 55 to 72%!

Connected vehicles is yet another large growth market as industries move quickly towards providing augmented and autonomous driving. A result of the Ericsson Mobility Report (2016) indicated that consumers expect reaction times on their devices to be below 6s which is part of the key performance indicators (KPI) of positive consumer experiences. The proceeding chapters will outline how the expansion of the analysis and design domain needs to be integrated with the creation of the next generation of architecture needed to support what 5G provides to individuals in every part of their lives.

1.6   Internet of Things (IoT)

To put IoT into perspective it is best to define it as an enabler to providing outcomes based on the collection of data. The objective for IoT can be thought of as a way to perfect a product faster. This means that new product releases can be achieved and vendors/businesses can get more immediate feedback and then adjust. It also means that it creates a more 24/7 analysis and design paradigm. Because data updates will be closer to real-time, products can meet what consumers tend to prefer—changes in consumer behavior and needs can be detected and modified in applications. In many ways, IoT creates a super intelligent monitoring system—a data aggregator combined with behavior activities.

IoT is built as a network stack made up of layers of interactive components. From a business perspective IoT possesses six essential analysis and design questions:

1. What software applications will reside on the device?
2. What hardware is best suited across the networks?
3. What data will be refreshed and sit on a device?
4. What are the external system interfaces?
5. What are the security considerations?
6. What are the performance requirements?

Figure 1.4 provides another view of these six questions.

IoT is built on the architecture that allows applications to reside across multiple networks. Exactly where these applications are located is part of the challenge of analysts. Specifically, IoT devices supported by the increased performance provided by 5G, will allow applications to execute on the device itself. This is currently known as Edge Computing, where devices will contain more software applications and data that can drive performance. Obviously, a program performing locally on a

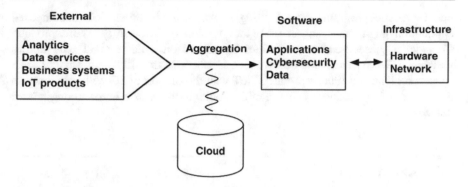

Components: Physical devices, sensors, data extraction, secured
communications, gateways, cloud, servers, analytics, dashboards

Fig. 1.4 Interactive components of IoT

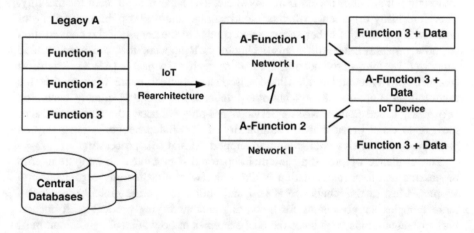

Fig. 1.5 IoT decomposition

device will outperform downloading the program and data from a remote server.
Resident programs and datasets then can be decomposed down to smaller units that
can perform the specific functions necessary on an independent and more autonomous device. This ultimately suggests that larger more complex legacy applications need to be re-architected into smaller component programs that can be
operated at the collection device as shown in Fig. 1.5.

We can see in Fig. 1.5 that a particular sub-function of the original Legacy A
application module is now decomposed to three subfunctions to maximize performance on IoT devices. I must emphasize that the critical increased capabilities to
design applications this way emanates from 5G's ability to transfer local data
among nodes back and forth more efficiently. It also increases speed to modify and

update mobile programs on the "Edge." An example of an IoT decomposed application might be a subset or lighter version of Microsoft's Word product. Consider a subset version that might be offered on a device that only allows viewing a Word document but without all the functionality. We already such subsets on IPad and IPhone products! IoT with the support of 5G will only increase these types of sub-versions because of the ability to move data faster among related devices.

1.7 Cloud

Cloud computing and IoT will develop yet another interesting combination of alternatives to where data resides and applications best perform. Obviously, cloud provides more operational performance and storage. Cloud has become the economical alternative to storing local applications and database storage; more importantly it provides access from anywhere. The latter is significant for mobility. There are many arguments whether cloud storage should be public or private or both, with the issue of cyber security and control at the center of the conversation on how organizations utilize this technology. It appears that the public cloud supported by third-party hosting companies such as Amazon (AWS), Microsoft (Azure), IBM Watson, Google Cloud, Cisco and Oracle to name a few, will be the predominant suppliers of the technology. Indeed, the Cloud is quickly becoming known as "Cloud Platform-As-A-Service." 5G only enhances the attractiveness of moving to cloud given that the complexity of distributed networks must rely on products and extensive data storage to support AI and ML processing.

The challenge of providing internal supported data centers to support interim processing and data manipulation is likely overwhelming for any organization to support. Most of this challenge is cost and ability to operate globally to support more complex supply chains for delivering and modifying product performance. Perhaps autonomous vehicles is the best examples of how 5G, IoT, and cloud must be able to reach almost every remote location thinkable to maximize consumer needs and services. Of course, the use of satellite technology makes most of this possible, but without the ability to add real-time performance and modification of data based on consumer behavior, connectivity has little attractiveness to providing point of contact operations.

From an analysis and design perspective, Cloud for a service is all about designing functional primitive applications. These primitive applications are essentially known as Application Program Interfaces or APIs that can be dynamically linked to piece together exceptional and agile applications. The Cloud providers will compete based on price of course, but also what APIs they make available that can easily provide development tools to help achieve quick program development. All of these Cloud providers thus present their own tool-kits of how one connects and builds these API products. The challenge for analysts and designers is to work with tool-kits that provide maximum transferability as it is likely that large organizations will choose to have multiple external cloud providers.

The prediction of expansion of IoT development dependent on Cloud is significant. According to Linthicum (2019), an EDC IoT study states that 55% of IoT developers connect through a Cloud interface, 32% connect via a middle tier, and 26% associate Cloud with IoT as a fundamental component. These stats will only increase as the IoT market is expected to reach 7.1 trillion dollars by 2020!

1.8 Blockchain

Blockchain represents the next major generation of systems architecture. Blockchain is really a data structure that builds on the concept of linked list connections. Each link or block contains the same transaction history. Thus, blocks can contain metadata—such as triggers, conditions, and business logic (rules) as well as stored procedures. Blocks can also contain different aspects of data. The design philosophy behind blockchain is that all blocks or nodes get updated when new transactions are made as data packages that must be accepted by all blocks in the chain. What is also significant about blockchain design is that access is based on key cryptography and digital signatures that will enhance security. The hope then is that blockchain provides the architecture that can maximize cyber security, especially of concern with the proliferation of IoT devices and wireless communication. The challenge with current blockchain architecture is latency consideration for time sensitive updating requirements, especially relevant for financial institutions.

Blockchain operates by appending new "blocks" to the chain structure. When data is part of any new transaction it becomes immutable and non-repudiated, that is, all valid transactions are added in real time updating. The blockchain has 5 properties:

1. *Immutability*: the events of an object cannot be changed, so that an audit trail of transactions is traceable.
2. *Non-repudiation*: the identity of the author of a transaction are guaranteed among all members of the blockchain.
3. *Data Integrity*: because of (1) and (2), data entry, manipulation and illegal modification are significantly reduced.
4. *Transparency*: all members or minors of the blockchain are aware of changes
5. *Equal Rights*: the rights can be set to be equal among all minors of the chain.

From the security perspective blockchain architecture offers the following features:

1. Because user or miner rights are set on the blockchain authorizations can be controlled. The fact that blockchains are distributed all members are dynamically informed of any changes.

2. The verification of any new member must be verified and self-contained, so
   invasions cannot come from outside or external systems. The verifier operates
   internally within the blockchain as a smart contractor and eliminates what is
   called "single points of failure" often relevant in decentralized network systems.
   Multiple verifiers can be enacted among integrated distributed networks along
   with arbitration software.

There are three current blockchain architectures: *Public, Consortium/
Community, and Private*. Public blockchains are essentially open systems accessible to anyone that has internet connectivity. Most digital financial currencies use
public blockchains because it provides better information transparency and audi-
bility. Unfortunately, the public design sacrifices performance as it heavily relies on
more encryption or cryptographic hash algorithms. Private blockchains are internal
designs that establish access for a specific group of participants that deal with a
particular business need. The Consortium/Community blockchain is a hybrid or
"semi-private" design. It is similar to a private blockchain but operates across a
wider group of independent constituents or organizations. In many ways a con-
sortium blockchain allows different entities to share common products and services.
In other words, it is a shared interest entity dealing with common needs among
independent groups.

The significant aspect of blockchain is that it is a ledger system. This means it
keeps information about the transaction—theoretically you could replay all the
transactions in a blockchain and should arrive at the same net results or disposition
of the data and its related activities. Blocks in the chain store information such as
date, time, and amounts of any transaction—like a purchase of goods. Further
blockchain stores information of who is participating in any transaction, so identity
of the individual or entity is recorded and must be known. Form a security perspective, blocks in the chain also store unique "hash" codes that act as a key to
access certain types of information and perform certain types of transactions.

From an analysis and design perspective it is important to first select the third
party blockchain product that will be used by the organization. Once this has been
assessed, there are a myriad of decisions on how the blockchain will operate both
administratively and configuration of the blockchain (data and computation).
Specifically, there are many data and computational decisions that need to be made
such as what will be stored in blocks, what administrative rights will be given to
different groups, and what general interfaces will be needed to access various types
of cloud stored data. Much of this will be covered in later chapters. Ultimately
blockchain represents a new type of distributed application architecture, a new kind
of client server model that is more peer-to-peer with embedded data and development applications. Analysts will need to understand traffic challenges in the
network and avoiding single points of failure, and determining API interfaces to
name a just a few design issues.

1.9 Cyber Security

Cyber security in analysis and design is perhaps the broadening dimension of change in designing hardware and software architectures. Cyber is the only component that is integral in every phase of the next generation. It simply is part of every analysis and design decision. Another way of articulating this point is to accept that cyber security must now be part of the analysis and design process! One must design for security to avoid creating Frankenstein 2.0. The first step to integrate cyber analysis and design is to accept its importance or to acknowledge and be aware of its importance. Analysts must navigate each component of this new generation of architecture and understand where exposures exist in their systems. Determining system exposures is now the most important part of analysis given that cyber protection is more a business decision than a design determination. Indeed, most cyber security professionals would state that almost any system can be designed to maximize security. Unfortunately, maximizing security will undoubtably limit performance to satisfy many consumer demands. The profession of cyber design then cannot exist without the existence of risk. Therefore, most security architectures must be part of any risk conversation. This then leads the analysts' job to interface with the necessary business and risk professionals in the organization. Some risk decisions are limited because of regulatory restrictions and legal limitations such as Europe's GDPR law. But many decisions must be exposed during the architectural design where someone along the way makes a decision on capability versus exposure. Thus, cyber security is a new component of the systems development life cycle (SDLC).

Another part of cyber design includes a number of literacy factors that exist in the internal community of any organization—what Gurak (2001) defined as cyberliteracy as a new internet consciousness. The fact is that cyber awareness is about the culture of the organization and analysts must gage the level of cyber sophistication of a population. Indeed, we know today that many breaches occur from careless behaviors of the employees of the organization. As a result, analysis must adapt a method of measuring cyber maturity of the organization and then factoring in risk as a fundamental part of designing new architectures and applications.

1.10 Quantum Computing

There is debate and uncertainty whether quantum computing could one day replace the silicon chip. Most quantum realities and applications still remain theoretical in nature. The functional component of quantum computers is called a qubit. A qubit is a complex and dimensional bit Binary computing, of course, is based solely on 0 and 1 sets of bits which perform one calculation at a time. In a qubit there are multiple 0's and 1's that can be performing simultaneous calculations and using multiple available resources to achieve output. Indeed, qubits might utilize atoms, ions, photons, or electrons differently even if the same calculation is repeated. In effect, qubits are like dynamic processing centers that use multiples of available parts to complete a task differently each time. While the resources can uniquely be determined upon each execution, there are mathematic probabilities that can be predicted under specific situations or states at the time a request is made. Obviously then a quantum computer could be a million times more powerful than today's computer.

Many of the details of the issues discussed thus far are covered in greater detail in forthcoming chapters. A brief description of each chapter follows.

Chapter 2: Merging Internal Users and Consumer Requirements

Analysts have traditionally focused on internal user requirements. These internal users were responsible for understanding what the business needed to support their customers. Depending on the business, a customer could be another business (B-to-B) or a consumer (B-to-C). This chapter addresses the need for the analyst to assess business requirements by going beyond the boundaries of the internal organization and learning how to work directly with outside customers and consumers. The next generation analyst must take on new challenges by providing ongoing and dynamic needs that mirror the uncertainties of their business environments. Therefore, analysts must create more speculative requirements based on market trends. The chapter will also address the need to integrate artificial intelligence and machine learning in the analysis process.

Chapter 3: Reviewing the Object Paradigm

Much of the development of IoT applications will require a significant proliferation of object-based reusable applications that will be replicated across complex networks and operate in mobile environments. This chapter will provide an understanding how object orientation works, methods of decomposing current larger products into functional primitives, and methods of determining where applications need to reside. This chapter also provides an overview of the tools or core concepts that analysts must use to create specifications, or the logical equivalences of what the system needs to do to provide answers for users. This includes completing the logical architecture of the system. Regardless of whether a package software system is required or the system is to be developed internally, the organization must create logical analysis before it can truly understand the needs of the business and its consumers. This chapter seeks to provide analysts with a path to transforming legacy systems into the new mobile-based paradigm of analysis and design.

Chapter 4: Distributed Client/Server and Data

This chapter covers how the back-end database engine is designed. The process of logic data modeling to produce a complete Entity Relational Diagram is covered, as well as methods of transferring data to multiple data storage facilities. Creation of data repositories is also discussed. The chapter effectively completes the data portion of the requirements document. In addition, the practice of normalization and denormalization are addressed and ways to determine the replication of data across mobile systems.

Chapter 5: The Impact of Wireless Communication

It is important to understand how 5G wireless affects application analysis and design. In order to assess this impact, it is necessary to review 5G's technical impact on performance. This Chapter discusses how the market will likely react to increased wireless performance and how 5G technology can be leveraged by application software developers. The chapter shows how the wireless revolution increases performance in a mobile environment, increases security and lowers latency. I also cover the evolving responsibilities of analysts as it relates to 5G architecture.

Chapter 6: The Internet of Things

The Internet of Things (IoT) represents the physical devices that will collect data for artificial intelligence and machine learning. The chapter shows how IoT represents the physical components that will make a technology feasible by placing intermediate smart hardware in every place imaginable around the globe. The chapter discusses how IoT will increase the uptime and real time processing as well as its ability to reduce unscheduled network failure. The chapter provides direction for the analyst with respect to security exposure and working with third-party vendors.

Chapter 7: Blockchain

This chapter provides an overview of the architecture of blockchain and its role in the mobile networks of the future. The chapter defines each type of blockchain, how it maximizes security using a ledger-based design. The chapter also identifies the advantages and disadvantages of blockchain and the way analysis and design should be conducted. Ultimately I will show how blockchain is required for expanding IoT and interfacing with cloud computing.

Chapter 8: Quantum Computing, AI, ML, and the Cloud

This chapter defines how quantum computing has the potential to change the processing capabilities of computing especially for ML and AI processing. The advantage of quantum is that it can do many calculations simultaneously and significantly reduce latency. Thus, quantum's role will be to crunch massive amounts of data to obtain valuable information that can be used to make predictions. The chapter will articulate how predictive analytics is quickly becoming automated using advanced AI APIs. These APIs will also be used to support more ML capabilities as the amount of data far surpasses a human's ability to do the analysis manually. The chapter presents the different approaches to ML and summarizes the advantages and disadvantages of using them. Of course, cloud processing becomes an important part of how best to distribute and process data across large distributed networks that are capturing data from mobile IoT devices.

Chapter 9: Cyber Security in Analysis and Design

This chapter shows how cyber security architecture requires integration with a firm's SDLC, particularly within steps that include strategic analysis and design, engineering, and operations in a distributed mobile environment. Mobile-driven applications need to comply with general standards for them to be useful, especially if they are integrating with legacy applications or part of a complete redesign using cloud. ISO 9000 is an international concept of standardization of quality that needs to be adopted in this transformation. This chapter covers many issues about cyber risks and the best practices that are currently be used and considered to combat the explosion of cyber-related crimes. The General Data Protection Regulation (GDPR) from the European Union is also discussed with recommendation on how the analyst can provide best practices.

Chapter 10: Transforming Legacy Systems

This chapter outlines the process of interfacing new mobile-based system with preexisting applications called *legacies*. Issues of product fulfillment, connectivity of legacy databases and processes, and integration of multiple systems architecture are covered. This chapter combines many of the suggested approaches to user interface and application specifications development that are covered in previous chapters. The objective of the chapter is to also provide a detailed pathway to ultimately converting legacy systems.

Chapter 11: Build Versus Buy

The classic question that has challenged business organizations is whether to build applications in-house or to buy them as pre-packaged software. This chapter will provide guidance on the appropriate steps to determine the right choice. Obviously, the decisions can be complex and vary depending on the application, the generic nature of the application itself, and the time requirements to have a functional application to meet business needs. Build decisions are further complex in that they can be developed in-house or an outsource provider. Buying can also come with choices on the amount of custom modifications that are necessary—the general rule being that over 20% modifications tend to be a bad choice for the buy equation. Cloud computing is the ultimate server-based paradigm to support IoT and blockchain technologies and we predict that most companies will engage in both development and third-party products.

Chapter 12: The Analyst and Project Management

This chapter provides guidance on system development life cycle methodologies and best practices for project management of in the next generation of systems. Project organization including roles and responsibilities are covered. There are many aspects of the next generation (5G, IoT, Blockchain) that are generic; however, there are certainly many unique aspects when managing these mobile-based systems. Thus, this chapter provides an understanding of where these unique challenges occur in the life cycle of software development. It also focuses on the ongoing support issues that must be addressed to attain best practices. The focus of the chapter is to suggest that analysts become involved in project management roles in addition to their responsibilities in the analysis and design of systems. The chapter also provides the necessary processes, recommended procedures, and reporting techniques that tend to support higher rates of project success. Furthermore, many projects have suffered because the management was not able to appropriately manage the contracted vendors. Organizations make the mistake of assuming that outsourced development and management is a safeguard for successful project completion. Organizations must understand that third party vendors are not a panacea for comfort and that rigorous management processes must be in place in order to ensure a successful project.

Chapter 13: Conclusions and the Road Forward

This chapter brings closure to the objectives of the book. At the forefront the chapter summarizes the technical and social architectures necessary to successfully transform into a digital organization. The analyst must sense opportunities and respond and understand the risk component as part of the analysis and design function. The chapter also defines the different types of generations at organizations and the importance of integrating Baby Boomer, Gen X, and Millennial (Gen Y) populations. Finally, the chapter seeks to emphasize the importance of the analyst roles and the myriad or roles and responsibilities that can be assumed to improve the probability of transforming organizations that will be driven by consumers in a mobile world.

1.11 Problems and Exercises

1. What are some of the limitations of traditional analysis and design?
2. What is meant by "consumerization of technology"?
3. How has the role of the analyst expanded in the digital era? Provide some examples.
4. Define and describe the components of the "New Paradigm." How does each of these components create more levels of complexity?
5. Discuss what is meant by the Mobile Connectivity Architecture. What are the common links?
6. Explain the relationships between the market and advances in technology.
7. What are the 6 essential analysis and design questions?
8. Explain IoT Decomposition.
9. What are the security advantages of blockchain?
10. How might quantum computing change the world?

References

Gurak, L. J. (2001). *Cyberliteracy: Navigating the internet with awareness*. New Haven, CT: Yale University Press.
Langer, A. M. (1997). *The art of analysis*. New York: Springer.
Langer, A. M. (2016). *Guide to software development: Designing & managing the life cycle* (2nd ed.). New York: Springer.
Langer, A. M. (2018). *Information technology & organizational learning: Managing behavioral change through technology and education* (3rd ed.). New York: CRC Press.
Linthicum, D. (2019). Talking to IoT Talk, TechBeacon. https://techbeacon.com/app-dev-testing/app-nirvana-when-internet-things-meets-api-economy.

Merging Internal Users and Consumer Requirements

2

This chapter seeks to provide analysts with a path to transforming legacy systems into the new mobile-based paradigm of analysis and design. In order to best understand this journey I must first clearly define what has been accomplished in the past; to do so provides today's analyst with a better understanding of why applications perform the way they were designed and the reasons why they are not capable of being used as we go forward with the new paradigm of advanced technologies in a mobile based global economy. Reviewing these methods also provides two other values, (1) it allows analysts to continue to support legacy applications and make enhancements to them until they are completely re-architected (which could take decades); and (2) not all the legacy analysis and design techniques should be eliminated, rather expanded to meet the needs of new digital based technologies.

So, the first part of this chapter will review existing methods and then expanding them for the newer generations of systems. The first aspect of understanding business requirements is the Tiers of Software Development.

2.1 The Tiers of Software Development

As stated, software development continues to evolve, particularly with the proliferation of internet-based wireless software products. The need to change the life-cycle of development certainly changes the way analysis and design is conducted. Unfortunately, many software products are created without thorough analysis and design, because it is easier just to create an "app" and then release it for consumer evaluation. Although this is an important development these advances in software development are overshadowing the importance of creating a parallel analysis and design paradigm.

© Springer Nature Switzerland AG 2020
A. M. Langer, *Analysis and Design of Next-Generation Software Architectures*,
https://doi.org/10.1007/978-3-030-36899-9_2

As the software industry focuses on integrated software solutions through robust mobile-based capabilities, it is important for the analyst to use the appropriate sequence of tiers to arrive at user and consumer requirements. Developers cannot expect good results from taking shortcuts, tempting as it may be to do so. The recommended sequence of tiers is outlined below.

2.1.1 User/Consumer Interface

Notwithstanding the type of software applications being developed, applications cannot be effectively designed without a user/consumer interface. The user/consumer interface tier acts as the base layer for any application because it drives the requirements of the product. Unfortunately, the user/consumer-interface is often bypassed because of pressures to issue product quickly. The *traditional SDLC* was most effective and often used during three fundamental phases:

1. Requirements analysis
2. Data Modeling
3. Normalization.

During requirements analysis, the development and design team conduct interviews in order to capture all the business needs as related to the proposed system. Data modeling involves the design of the logical data model which will eventually be transformed into a physical database. Normalization is conducted to reduce the existence of redundant data. Below is a more specific depiction of the Development, Testing, and Production cycles of the SDLC.

1. Development

The Development life cycle consists of all the necessary steps to accomplish the creation of the application. The four components are feasibility, analysis, design, and the actual coding. The feasibility process helps determine whether the application is realistic and has an acceptable return-on-investment (ROI). ROI usually has complex financial models that calculate whether the investment will provide an acceptable rate of return to the business. ROI should not solely use monetary returns as the only method; there are a number of reasons why companies develop software solutions that are not based on monetary returns (Langer 2011a, b). Feasibility reports typically contain ranges of best and worst cases. Feasibility also addresses whether the business feels can deliver on time and on budget.

Analysis is the phase that delivers a logical requirements document. Indeed, the analyst creates the blueprint for programmers and database developers. Analysis, as an architectural responsibility is very much based on a mathematical progression of predictable steps. These steps are quite iterative in nature, which requires practitioners to understand the gradual nature of completion of this vital step in Development. Another aspect of the mathematics of analysis is decomposition.

Decomposition as we will see establishes the creation of the smaller components that make-up the whole. It is like the components of a human body that when put together makes up the actual person that we physically see. Once a system is decomposed, the analyst can be confident that the "parts" that comprise the whole is identified and can be reused throughout the system as necessary. These decomposed parts are called "objects" and comprises the study and application of object-oriented analysis and design. This traditional approach, actually is the key to moving forward to providing reusable mobile-based applications. Therefore, the basis of an effective path is whether the legacy system has been decomposed to the object level. Unfortunately, most major legacy systems are not yet in this state. So the first step in a transition is to move them into reusable parts, just like those found in an automobile—tires that can fit many different vehicles.

The design step while less logical is much more a more creative phase. Design requires the analyst to make the physical decisions about the system, from what programming language to use, which vendor database to select (Oracle, Sybase, DB2 for example), to how screens and reports will be identified. The design phase can also include decisions about hardware and network communications or the *topology*. Unlike analysis, design requires less of a mathematical and engineering focus, to one that actually serves the user or consumer view. The design is often more iterative, which could require multiple sessions with users and consumers using a trial and error approach until the correct user interface and product selection has been completed. We will see that the new paradigm requires much more design and physical trials than getting it right just in analysis. While this sounds a bit strange, we will see that many applications are developed, tested as well as they can, but then once put into consumers use, they often need a lot of changes. This is where the consumer interface has significantly changed the way we work on the SDLC.

Actual coding represents another architectural as well as mathematical approach. However, while early programming languages were very close to the machine, they are now several layers back or what we call abstractions of what the actual code that the machine understands. That is, software is the physical abstraction that allows us to talk with the hardware machine. Coding then is the best way to actually develop the structure of the program. Much has been written about coding styles and formats. The best known is called "structured" programming. Structured programming was originally developed so that programmers would create code that would be cohesive, that is, would be self-reliant. Self-reliance in coding means that the program is self-contained because all of the logic relating to its tasks is within the program. The opposite of cohesion is coupling. Coupling is the logic of programs that are reliant on each other, meaning that a change to one program necessitates a change in another program. Coupling is viewed as being dangerous from a maintenance and quality perspective simply because changes cause problems in other reliant or "coupled" systems. The relationship to coding to analysis can be critical given that the decision on what code will comprise a module may be determined during analysis as opposed to coding. Today, software programming languages allow less "technically" trained people to use them, and this has allowed for a larger and growing number of professionals that are developing products.

2. Testing

Testing can have a number of components. The first form of testing is called program debugging. Debugging is the process where a programmer ensures the application executes. For this reason, we consider debugging part of the programmer's responsibility. This is very different than a formal testing or quality assurance group of staff. The challenge is always who does what and when is a program ready for the quality assurance group to ensure that the program delivers the behavior and outputs originated from the requirements document. Programmer's then should never pass a program to quality assurance that does not execute, or at least executing properly under all conditions.

The formal process should be that a "debugged" program should be forwarded to a formal quality assurance group for validation. Most IT organizations have developed formal QA departments that are comprised of non-programmers. These QA groups focus on testing the correctness and accuracy of programs. Quality assurance organizations typically accomplish this by designing what is known as Acceptance Test Planning. Acceptance Test Plans are designed from the original requirements, which allow quality assurance personnel to develop assurance testing based on the user's original requirements as opposed to what might have been interpreted. For this reason, Acceptance Test Planning is typically implemented during the analysis and design phases of the life cycle but executed during the Testing phase. Acceptance Test Planning also includes system type testing activities such as stress and load checking that ensures that the application can handle larger demands of data, consistent access or number of users on the system simultaneously. It also addresses compatibility testing, such as ensuring that applications operate on types of browsers or computer systems. QA of course is an iterative process that can often create iterations of redesign and programming. The acceptance testing has two distinct components: (1) the design of the test plans, and (2), the execution of those acceptance plans. In the "mobile age" of software development, it is necessary to have the programming and testing process be happening more simultaneously. This can occur because of the ability to make changes quicker and identify problems because of the decomposition of the smaller program functions.

3. Production

Production is really the "going-live' phases. Ultimately, Production needs to ensure the successful execution of all aspects of a system. During Production, there is the need to establish how problems will be serviced, what support staff will be available and when and how inquiries will be responded to and scheduled for fixing. This component of Production may initiate new Development and Testing cycles because of redesign needs (or misinterpreted user needs). This means that the original requirements were not properly translated into system realities. However, today's systems are more like living organisms that are always evolving, always providing new capabilities, always in testing, and always going into product.

There are other aspects of Production as a Life Cycle includes that have not changed:

- Backup, recovery, and archival;
- Change control;
- Performance fine-tuning and statistics;
- Audit and new requirements.

2.1.2 Tools

Software systems require that analysts have the appropriate tools to do their job, just like an architect. Many new techniques are needed both in the short-term and the long-term of the analyst profession. Furthermore, and an even more significant challenge, is understanding which of the many available tools to use at any given point. Analyst tools are often designed for specialized use rather than for general application, and using the wrong tool can potentially cause significant damage. Finally, the sequence of use for each specialized tool is also critical to success. Indeed, the order of operation, as well as the relationship among specialized analysis tools, must be mastered to ensure success. The newer tools discussed herein will obviously need to target the wireless and mobile needs of consumers.

2.1.3 Productivity Through Automation

Having the appropriate tools and knowing how and when to use them is only part of the formula for success. Analysts must also be productive—and productivity can be accomplished only through the use of automation. Automation is implemented using integrated various automated products or what was once defined as Computer Aided Software Engineering or CASE. These products provide the analyst with an automated and integrated toolsets that are centralized through a core automated system and repository of data definitions to be used by all products in the info-systems of an enterprise.

2.1.4 Object Orientation

Perhaps the most important tool for the wireless generation of software products is the concept of object orientation (OO). Whether or not software systems are OO compliant, analyzing systems using the object method builds is essential for created functional primitive objects that can be disseminated across IoT devices. OO developed software creates better systems that are more cohesive, reusable, and maintainable. Such code is more maintainable and the foundation of the development of reusable components that can be integrated across architectures and combined dynamically into larger applications. Without an OO design philosophy, systems tend to have parts in many applications that are re-coded and virtually impossible to maintain. Welcome to the legacy challenge! With the advent of mobile-based architectures it is vital to convert all legacy software into an object library repository. The question that this book addresses is, how?

2.1.5 Client/Server

In many ways a large portion of legacy software is still governed by the concept of client/server processing. Client/server design was born out of a master/slave philosophy, where the server contained the major code and the databases, and the client had local needs, mostly to help with performance. Client/server is now outdated and must be replaced with a network strategy of linked components that may, or may not need a master server, rather a more flattened linkages of parts. Thus, client/server software development was originally designed to solve a network performance problem, but 5G and future quantum type hardware will simply continue to make that architecture obsolete. with network hardware strategy. While client/server hardware topology is an important issue in itself, it has little to do with the process of deciding how software modules should interact across the network and where such modules should be placed. Such decisions must be driven by issues that arise during the process of analysis. Client/server software processing, in its true implementation, involves the interaction of objects and defining the way in which they will communicate with each other across IoT devices. The network will simply act as connection points. Thus, analysts must first be versed in the laws governing OO if they are to understand how to design mobile-based IoT solutions.

2.1.6 Internet/Intranet to Mobility

The movement to cyber communication across the internet with web-based technologies, was initially coined Internet/Intranet processing. It resulted in the introduction of a new breed of software applications. These new applications certainly brought new challenges to analysts and designers. Increasingly, analysts themselves had to work directly with commercial advertisers and marketing departments to create a new "look and feel" that were demanded by consumers using the internet to access products. These web-based systems, inserted the analyst into a new part of the development process, no longer just limited to gathering requirements. The analyst now in the wireless IoT era is now the critical integrator to transform the systems. We have seen less and less distribution of development teams, with the coming of cloud companies that can develop object modules easier and store them for distribution across complex networks that feed IoT devices, that is, companies will find more and more outsourced solutions to fill their needs. So, the term Internet/Intranet is no longer relevant—it is now Mobility!

Table 2.1 Tiers of analysis and software application development

Tier	Analyst application
6	Mobility and IoT
5	Distributed networks—breaking down applications
4	Object orientation—selection of objects and classes
3	CASE—automation and productivity of Tier 2
2	Structured tools—Use cases, DFD, PFD, ERD, STD, process specification, data repository
1	User/Consumer interface—interviewing skills, Marketing, Risk Analysis

Mobile-based processing requires that analysts to have mastered the client/server paradigm more as a distributed network of parts. Indeed, many professionals will dub mobility development as "client/server grown-up." This may not be the best definition of architectural agility, but it functionally supports the tier concept of dynamic and connected parts.

So the new Tiers of Software Development that I developed in 2011 is now mobile IoT. I call each of these "tiers" because of their dependence on the previous phase as a building-block nature and their inevitable dependence on each other. I insist that effective analysts must master these tiers to ensure success at the next phase. I present these tiers in Table 2.1.

The table graphically shows how each tier must be dependent on the other. There is a profound message in this diagram which suggests that no tier can be developed or exist without the previous one. To ensure success on a project, everyone involved in the design and development of application software must fully understand the interdependent nature of these tiers. Analysts must be able to convey to their colleagues that to go mobile and IoT, organizations must first have excellent user/consumer interfaces, mastery of a structured toolset, a vehicle for automation so that the process will be productive, an understanding of the concept of objects, and a way to deploy these objects in a distributed outsourced cloud environment.

The following sections provide a step-by-step process of gathering the data that you need to create a *traditional* requirements document.

2.2   Establishing Internal User Interfaces

The success factors in analysis start with the established interfaces from day one. What does this mean? You must start the process by meeting with the right people in the organization. In the best projects, the process is as follows:

1. Executive Interface: There needs to be an executive-level supporter of the project. Without such a supporter, you risk not being able to keep the project on schedule. Most important, you need a supporter for the political issues that you may need to handle during the project (discussed in detail later). The executive

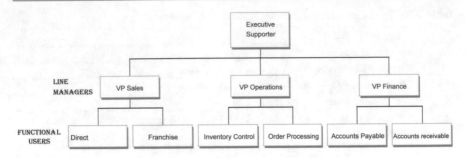

Fig. 2.1 Established interface layers

supporter, sometimes known as a Sponsor should provide a preliminary
schedule advising the organization of what is expected and the objectives of the
project. The executive supporter should attach a letter to the preliminary
schedule and send it to the project team members. The letter must put the
importance of the project into perspective. Therefore, it is strongly recommended that you draft this letter yourself or at least have influence over its
content, since doing so can ensure that the message is delivered appropriately.
The executive supporter should also establish regular reviews with the analyst
and the user community to ensure that objectives are being met.
2. Department Head or Line Manager Interface: If appropriate, the Department
   Head should provide guidance about which individuals should represent the
   department needs. If there are several people involved, the analyst should
   consider a JAD-like approach. Depending on the size of the organization, the
   Department Head might also establish review sessions to ensure compliance.
3. Functional User Interface: Perhaps the most important people are the ones who
   can provide the step-by-step needs of the system. Figure 2.1 shows a typical
   organization interface structure.

2.3   Forming an Interview Approach

The primary mission of an Analyst or Systems Designer is to extract the physical
requirements of the users and convert each to its Logical Equivalent. The most
critical step in this mission is the actual interview, in which you must establish a
rapport with the user(s) that will facilitate your obtaining the information you need.
Your approach will dramatically change based on the level and category of the
individual being interviewed. Therefore, prior to meeting with any user, it is critical
to understand the culture of the company, its past experiences with automation and
most important its organization structure.

The following five-step procedure will help guide you more smoothly through
the interview process.

Step 1—Get The Organization Chart:

There are few things that are more useful in understanding the chain of command and areas of responsibility. Depending on the size of the enterprise, and the scope of the project, the Organization Chart should start at the Executive Supporter level and work down to the Operational Users.

Step 2—Understand Everyone's Role in the Organization Chart:

If there are any individuals not involved in the project (who should be, given their position in the organization), first ask why, then make a notation for yourself that they are not to be included. Management may assume an individual or role should not be included and may often overlook their importance. Do not be afraid to ask why a person is not deemed necessary for the analysis of the system, and determine if you are satisfied with the reasons for their exclusion. Remember, you can still control and change the approach at this point, and management will probably respect you for doing so.

Step 3—Assume the Situation is Political:

Be sure you understand the personalities that you will have to deal with. In almost any implementation, politics among people becomes part of the process. To ignore its existence—and the constraints it is likely to impose—is to invite failure. The question is how to obtain information about internal politics. The best approach is to start as high up in the organization as possible, typically at the executive supporter level. You might be surprised at the amount of information they have. Of course, you should not ask explicitly about the politics, but rather phrase your question as follows: "Can you give me some perspective on potential department and personnel conflicts that may occur during the interview cycle and that I should be aware of?" You may not always get the answer you need, but if you keep asking the question during every interview, you will discover a great deal about the way the organization functions. And remember, only people make projects complex!

Step 4—Obtain Information about User Skill Sets:

To start an interview without knowledge of the user's technical skills puts the analyst at a huge disadvantage. Having this information will allow you to formulate a plan of questions and to determine the best approach to the interview. If the user has no knowledge, the questions should be tailored to include a minimum of technical content. The following guidelines for preparing for interviews reflect a common-sense approach, yet it is amazing how many analysts fail even to consider such strategies!

1. Gather information before the session to allow the user—as well as yourself—to be prepared and to give you both a much clearer understanding of what will be covered during the interview.
2. Develop a questionnaire. Technical questions should be phrased differently depending on the level of knowledge possessed by the user.

3. Determine whether the interview will provide enough input to obtain the necessary information. This is not always the case; however, it happens more often than you might think. Understanding user capabilities before the interview may change not only the scope of the meeting, but may also suggest who, in addition to the user, may need to be in attendance at the interview.

Step 5: Arrange for a Pre-meeting with the User:

A pre-meeting may not always be possible, and in any case it must be a short meeting, perhaps half an hour. The session should be designed to be high-level and provide a general idea of what will be covered during the actual interview. But more important, it will allow you to get a snap-shot of the user. You might say you are obtaining a "comfort level" (or "discomfort level") for that user, and such meetings can provide you with an idea of what to expect and how to finalize your approach. What do you look for? Here is some direction:

1. The pre-meeting should give you enough feedback to place or confirm the user's technical level.
2. Look at everything in the user's office or their environment. Is it sloppy? Is it tidy and organized? The state of the user's environment will often be consistent with the way they provide information. The insight you gain from observing the environment should give you guidance about the types of questions to ask this individual.
3. Look for signs of attitude. The user's level of interest should be evident. Do they view the upcoming session as a waste of time, or are they excited about the meeting?
4. The information gleaned in the pre-meeting can provide you with helpful hints about what to expect from the interview and from the user in general.

2.4   Dealing with Political Factions

The importance of internal politics at the user's site should never be underestimated. Perhaps the most common question raised by both professionals and student analysts is how to provide quality analysis when office politics get in the way. Here are some guidelines:

1. First, assess whether you are in the No-Win Scenario. Many of us hate to admit that the No-Win Scenario does indeed exist in many environments, but you should be on the lookout for the signs. If your manager will not support you, if the company is underpaying you, if the users hate you, if there are no automated tools to do the analysis, and if upper management doesn't care, then you are in a difficult position. If you cannot change the situation, you must inform management that the results of your analysis will be significantly impaired by the

lack of support and tools to complete the project properly. The techniques offered in this book assume that all parties are interested in providing the best solution possible, not in providing a system that is barely adequate.

2. On the other hand, do not be too quick to assume that you are in the No-Win Scenario. Most politically hampered projects need some strategy to get them on course, and most problems can be overcome if you know how to approach them. Here is a typical example of such a problem and some ideas you can apply to solve it.

Problem:

The Users who currently operate the system won't talk to me. They are afraid either that the new system might replace them or that their jobs will significantly change. In short, they fear change.

Recommended Solution:

Most operational users are managed by a Supervisor or "In-Charge." Sometimes even a Line Manager can be directly responsible for production workers. In any event, you must determine who is responsible and meet with that person. The purpose of the meeting is to gain their support. This support is significant since you might find that the Supervisor was once in operations and will be able to understand the problems you may encounter. If the meeting is successful, the Supervisor may be able to offer a strategy. This strategy can vary from a general meeting with the users, to individual discipline, to escalation to upper management. Whatever you do, do not allow such a situation to continue and do not accept abuse; to do so will ultimately reflect on you and your abilities.

Obviously, if the Supervisor is also a problem, then you have no choice but to go to upper management. However, this option is not a desirable one from the analyst's viewpoint. Upper management's reaction may not be helpful, and it could be damaging. For example, they might be indifferent to your problem and instruct you to deal with it yourself, or they might simply send the Supervisor a letter. In some cases you may be fortunate and the Supervisor's responsibilities regarding the system will be given to another manager. Consider, though, how unpleasant the consequences may be if you appeal to upper management and get no support: you may be left working with an already-unhelpful Supervisor who has been made even more so by your complaint. It is important to remember that once you go to upper management, the line has been drawn. Supervisors typically are responsible for the day-to-day operation. They usually know more about the entire operation than anyone else, and therefore you are well advised to find a way to get them on your side. A supportive Supervisor can be invaluable in helping you overcome problems, as long as you are not shy about suggesting ways to get the users comfortable.

2.5 Categories and Levels of Internal Users

Establishing user interfaces represents the vehicle to formulate much of the interview approach. It is necessary; however, to go further into the characteristics of the people particularly with respect to the *category* and *level* they have within the organization. Figure 2.1 established the three general categories, called executive, department head or line manager, or functional. It is important to explore their characteristics. In order that we better understand each category, I have always asked the following question: *What would be their interest in the success of the project, that is, what would make them happy with the new system?* Let's apply this question for each user category.

1. *Executive users*: individuals at this layer are most interested in the concept of return-on-investment (ROI). ROI basically focuses on whether an investment will provide a financial return that makes the effort worthwhile to the organization. While there are many comprehensive formulas that are often applied to the study of ROI, our context pertains to the short- and long-term benefits of investing in building new software. There are generally five reasons why executives agree to fund software development. They are listed in order of significance to the investor.

   a. *Monetary return*: simply put this means that the software will generate dollar revenue. An example might be the Internet software that supports on-line ordering systems such as Amazon has for book shipments. Their system not only provides the functionality to handle shipments, but provides a web interface that can be directly associated with revenues provided by book orders through the Internet.
   b. *Increased productivity*: many software systems are unable to demonstrate direct monetary benefits; however, many of them are developed to increase productivity. This means that the system will allow organizations to actually produce and deliver more. Thus, the system allows the organization to derive higher revenues through increased productivity of its resources.
   c. *Reducing costs*: software projects are approved so that organizations can reduce their existing overhead costs. This typically relates to the replacement of manual activities with computer ones. While reducing costs appears to be similar in nature to increasing productivity, they are often implemented for different reasons. Increased productivity usually relates to organizations that are growing and are looking for ways to improve output because of very high demand. Reducing costs, on the other hand, can represent a defensive measure, where an organization is seeking to find ways to cut costs because of a shrinking market.
   d. *Competition*: software systems are created because the competition has done so. Therefore, competitive reasons for producing software is a defensive measure against someone else who has demonstrated its value. An example of this is in the banking sector. Citibank was one of the first banks to

introduce automated teller machines (ATM). Other banks soon followed because of the success that Citibank had with proliferating ATMs throughout New York State. This does not imply; however, that competitive systems are always defense mechanisms, indeed, many commercial web sites are being introduced based simply on market forecasts for their potential to increase business.

e. *For the sake of technology*: while not the most popular, some organizations will invest in new systems because they think it's time to do so or they are concerned that their technology is getting old. This way of supporting new systems development is rare, as it suggests the spending of money without a clear understanding of its benefits.

Therefore, the executive category of users is one that is interested in the value of the investment. These users have a global view of needs as opposed to the details. in fact, they may know little about how things are really done. The value of the executive interface is to provide the scope and objectives of the project against their perceived value they intend to get from the software. Another popular phrase for this is called the *domain* of the system. Domain often refers to *boundaries*. Ultimately, what makes them happy is a system that delivers what was promised or the expected ROI.

2. *Department head or line manager users*: these users represent two main areas of user input. First they are responsible for the day-to-day productivity of their respective departments. Thus, they understand the importance of meeting the objectives of the organization as set forth by the executives. Indeed, they often report the executives. On the other hand, department heads and line managers are responsible to their staff. They must deal with the functional users and prescribe ways to improve both their output and their job satisfaction. These users perhaps provide what I call *the best bang for the buck*, a phrase that usually means that for the time, you get the most. One can see that the department heads and line managers are responsible for most of what happens every day in an organization. Another phrase that can be used to describe them is your *Most Valuable Players (MVPs)*. However, beware, MVPs are the hardest to find and get for the interviews. What makes department heads and line managers happy is the most complex. They want a systems that produces the output that they are expected to provide and they need a system that makes keeps their staff happy and productive.

3. *Functional users*: also known as the users in the trenches, these people essentially do the operational activities. While they know a lot about their processes, they usually care little about the productivity and expected ROI. I often see these users as people who want little pain, and just want to work the hours they need to. Thus, fancy systems are of little interest to them unless they provide no pain —and no pain derives to a system that makes their job easier.

The next area to understand about users is their level. By level, I mean their understanding of computers. There are three levels of users:

1. *Knowledgeable*: the determination of knowledge can be tricky and certainly based on someone's opinion. I define knowledge in reference to experience. An experienced user can be defined as a person who "has been through it before." A user who has been through the development of a new system can therefore be defined as "knowledgeable" within this context.
2. *Amateur*: the definition of an amateur is based not so much on experience, but rather to the type of experience the user has. Amateurs can be thought of as hobbyists who enjoy working with computers at home, but have no professional experience in developing software in an organization. In this perspective, I believe the meaning an amateur is globally defined as one who does not get paid for the work.
3. *Novice*: these users have no experience with computers. While there are fewer of such users than 10 years ago, they still exist. A better way of perceiving a novice user is to relate to my definition of knowledgeable. In this context, a novice user is one that has never been part of the implementation of a new system in a professional environment.

Perhaps the most problematic of the above levels is the amateur. I have found that users who are knowledgeable provide benefit to projects. They, in many ways, act as a checkpoint for the analyst in that they can ask good questions and particularly remember historical problems that actually can help the development process. Novice users add little value and also add few problems. They tend to do what you ask of them. Amateurs, on the other hand, tend to know enough to be dangerous. They also tend to have a profound interest in the topic that they often get-off on tangents about the technology as opposed to concentrating on the particulars of the project.

What is most important is the mapping of these categories and levels. An analyst might interview a knowledgeable executive, or a novice functional user. Each permutation can affect the way interviews are conducted. For example, an interview with a group of amateurs would focus the analyst on ensuring that the agenda is very specific, otherwise, discussions could easily get-off track. Therefore, the understanding about user levels and categories can only assist in the development of effective interview approaches.

2.6 Requirements Without Users and Without Input

Could it be possible to develop requirements for a system without user input or even consumer opinions? Could this be accomplished?

Perhaps we need to take a step back historically and think about trends that have changed the competitive landscape. Digital transformation is may indeed be the most powerful agent of change in the history of business.

We have seen large companies lose their edge. IBM's fall as the leading technology firm in the 1990s is an excellent example, when Microsoft overtook them into that position. Yet Google was able to take the lead away from Microsoft in the area of consumer computing. And what about the comeback of Apple with its new array of products? The question is: Why and how does this happen so quickly?

Technology continues to generate change, or as it is now referred to as "disruption." The reality is that it is getting increasingly difficult to predict what consumers want and need—if they even know! The challenge then is how can we forecast the changes that are brought about by technology disruptions? So the digital transformation is more about predicting consumer behavior and providing new products and services, which we hope consumers will use. This is a significant challenge for the analyst of course, given that the profession was built on the notion that good specifications accurately depicted what users wanted. Langer (1997) originally defined this a the "Concept of the Logical Equivalent." So we may have created an oxymoron—how do we develop systems that the user cannot specify? Furthermore, specifications that depict consumer behavior is now further complicated by the globalization of business. Which consumer behavior are we attempting to satisfy and across what cultural norms?

So the reality is that new applications will need more generic and be built with a certain amount of risk and uncertainty. That is business rules may be more questionable, risks will need to be evaluated and consistent with those practiced by the organization.

Let me state a case for my argument of designing systems for uncertainty and change.

If we look at the successful application in the 1980s with the advent of the personal computer, a standout is the electronic spreadsheet. The electronic spreadsheet was first introduced by a company called Visicorp and named the product VisiCalc. It was designed for the Apple II and eventually the IBM PC microcomputer. The electronic spreadsheet was not built off of consumer needs, rather perceived needs. The electronic spreadsheet was designed as a generic calculator and mathematical worksheet. Visicorp took a risk by providing a product to the market and hoped that market would find it useful. Of course history shows that it was a very good risk. The electronic spreadsheet, which is now dominated by the Microsoft Excel product have gone through multiple generations. The inventors had a vision and the market then matured its many uses. Thus Visicorp was correct, but not 100% accurate with what consumers would want and continue to need. For example, the additional feature of a database interface; three-dimensional spreadsheets to support budgeting; and forward referencing are all examples of responses from consumers that generated new enhancements to electronic spreadsheets.

Another useful approach to dealing with consumer preferences is Porter's Five Forces Framework. Porters framework consists of the following five components:

1. *Competitors*: what is the number of competitors in the market and what is the organization's position within the market?
2. *New Entrants*: what companies can come into the organization's space and provide competition?
3. *Substitutes*: what products or services can replace what you do?
4. *Buyers*: what alternatives do buyers have? How close and tight is the relationship between the buyer and seller?
5. *Suppliers*: What is the number of suppliers that are available which can affect the relationship with the buyer and also determine price levels?

Porter's framework is graphically depicted in Fig. 2.2 (Table 2.2).

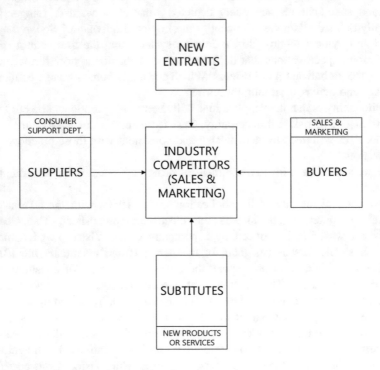

Fig. 2.2 Porter's five forces framework

Table 2.2 Langer's analysis consumer methods

Porter's five forces	Cadel et al.'s approach	Langer's sources of input
Industry competitors	How strong is your market share	Third party market studies
New entrants	New threats	Third party market studies
		Surveys and focus groups
Suppliers	Price sensitivity and closeness of relationship	Consumer support and end user departments
Buyers	Alternative choices and brand equity	Sales/Marketing team
Substitutes	Consumer alternatives	surveys and focus groups
		Sales and marketing team
		Third party studies

2.6.1   Concepts of the S-Curve and Digital Transformation Analysis and Design

Digital transformation will also be associated with the behavior of the S-Curve. The S-Curve has been a long-standing economic graph that depicts the life cycle of a product or service. The S-Curve is just that as shown below in Fig. 2.3.

At the left lower portion of the S-Curve represents a growing market opportunity that is likely volatile and where demand exceeds supply. Therefore, the market opportunity is large with high prices for the product. Thus, businesses attempt to capture as much of the market at this time, which in turn requires risk-taking with associated rewards, especially in increasing market share. The shape of the S-Curve suggests the life of the opportunity.

As the market approaches the middle of the center of the S, demand begins to equal supply, prices start to drop and the market, in general, becomes less volatile and more predictable. The drop in price reflects the presence of more competitors.

Fig. 2.3 The S-Curve

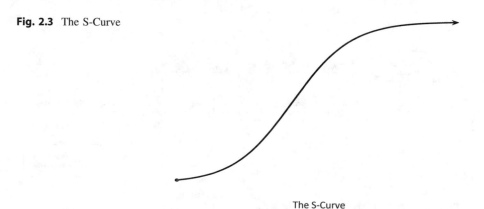

The S-Curve

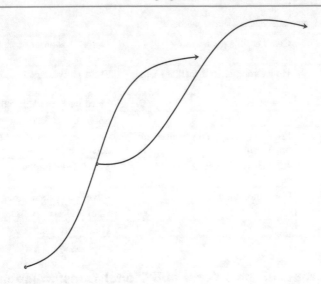

Fig. 2.4 Extended S-Curve

As a product or service approaches the top of the S, supply begins to exceed demand. Prices begin to fall and the market is said to have reached maturity. The uniqueness of the product or service is now approaching commodity. Typically, suppliers will attempt to produce new versions to extend the life of the curve as shown Fig. 2.4.

Establishing a new S-Curve then extends the competitive life of the product or service. Once the top of the S-Curve is reached, the product or service has reached the commodity level, where supply is much greater than demand. Here, the product or service has likely reached the end of its useful competitive life and should either be replaced with a new solution or considered for outsourcing to a third-party.

Langer's Driver/Supporter depicts the life cycle of any application product as shown in Fig. 2.5.

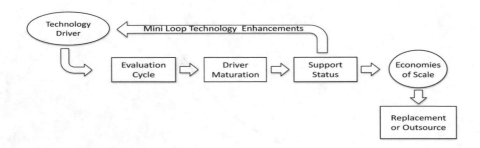

Fig. 2.5 Langer's drive/supporter life cycle

Table 2.3 S-Curve, application requirement sources, and risk

S-Curve Status	Analysis input source	Risk factor
Early S-Curve	Consumer	High, market volatility and uncertainty
High S-Curve	Consumer	Lower, market is less uncertain as product becomes more mature
	End users	Medium, business users have experience with consumers and can provide reasonable requirements
Crest of the S-Curve	End users	Low, business users have more experience as product becomes mature
	Consumer	High, might consider new features and functions to keep product more competitive. Attempt to establish new S-Curve
End of S-Curve	End user	None, seek to replace product or consider third-party product to replace what is now a legacy application. Also think of outsourcing application

2.7 Analysis and Design and the S-Curve

When designing a new application or system, the status of that product's S-Curve should be carefully correlated to the source of the requirements. Table 2.3 reflects the corresponding market sources and associated risk factors relating to the dependability of requirements based on state of the consumer's market.

2.8 Communities of Practice

Another technique that can be used to obtain more accurate information in the digital economy is called Communities of Practice (COP). COP has been traditionally used as a method of bringing together people in organizations with similar talents, responsibilities and/or interests. Such communities can be effectively used to obtain valuable information about the way things work and what is required to run business operations. Much of this information is typically implicit knowledge that exists undocumented in the organization. Getting such information strongly correlates to the challenges of obtaining dependable information from the consumer market. I discussed the use of surveys and focus groups earlier in this section, but COP is an alternative approach to bringing together similar types of consumers grouped by their interests and needs. Communities of practice are based on the assumption that learning starts with engagement in social practice and that this practice is the fundamental construct by which individuals learn (Wenger 1998). Thus, COPs are formed to get things done by using a shared way of pursuing interests from common users. For analysts this means another way of obtaining requirements by engaging in, and contributing to, the practices of specific consumer communities. This means that working with COP is another way of developing relations with consumers to better understand their needs. Using this approach inside an organization provides a means of better learning about issues by using a sustained method of remaining interconnected with specific business user groups, which can define what the organization really knows and contributes to the business that is typically not documented. The notion of COP supports the idea that learning is an "inevitable part of participating in social life and practice" (Elkjaer 1999: 75). Thus, analysts need to become engaged in learning if they are to truly understand what is needed to develop more effective and accurate software applications. Communities-of-practice also includes assisting members of the community, with the particular focus on improving their skills. This is also known as "situated learning." Thus, communities-of-practice is very much a social learning theory as opposed to one that is based solely on the individual. Communities-of-practice has been called learning-in-working where learning is an inevitable part of working together in a social setting. Much of this concept implies that learning in some form or other will occur and that it is accomplished within a framework of social participation, not solely or simply in the individual mind. In a world that is changing significantly due to technological innovations, we should recognize the need for organizations, communities, and individuals to embrace the complexities of being interconnected at an accelerated pace.

There is much that is useful in communities-of-practice theory and that justifies its use in the analysis and design process. While so much of learning technology is event-driven and individually learned, it would be short-sited to believe that it is the only way learning can occur in an organization. Furthermore, the enormity and complexity of technology requires a community focus. This would be especially useful within the confines of specific departments that are in need of understanding how to deal with technological dynamism. That is, preparation for using new technologies cannot be accomplished by waiting for an event to occur, instead, preparation can be accomplished by creating a community that can assess technologies as a part of the organization's normal activities. Specifically this means that through the infrastructure of a community, individuals can determine how they will organize themselves to operate with emerging technologies, what education they will need, and what potential strategic integration they will need to prepare for changes brought on by technology. Action in this context can be viewed as a continuous process, much in the same way that I have presented technology as an ongoing accelerating variable. However, Elkjaer (1999) argues that the continuous process cannot exist without individual interaction. As he states:

> Both individual and collective activities are grounded in the past, the present, and the future. Actions and interactions take place between and among group members and should not be viewed merely as the actions and interactions of individuals (p. 82).

Based on this perspective, technology can be handled by the actions (community) and consumers (individuals) as shown in Fig. 2.6.

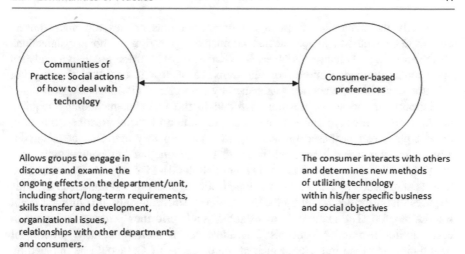

Fig. 2.6 Community and consumer actions

It seems logical that COP can provide the mechanism to assist analysts with an understanding of how business users and consumers behave and interact. Indeed, the analyst can target the behavior of the community and its need to consider what new organizational structures can better support emerging technologies. I have in many ways already established and presented the what should be called the "community of IT Analyst practice" and its need to understand how to restructure in order to meet the needs of the digital economy. This new era does not lend itself to the traditional approaches to analysis, but rather to a more community-based process that can deal with the realignment of business operations integrated with different consumer relationships.

Essentially, "communities of IT Analyst practice" must allow for the continuous evolution of risk-based analysis and design based on emergent uncertain strategies. Emergent uncertain strategies acknowledge unplanned actions and evolutions in consumer behavior, which have been historically defined as patterns that develop in the absence of intentions (Mintzberg and Waters 1985). Emergent uncertain strategies can be used to gather groups that can focus on issues not based on previous plans. These strategies can be thought of as creative approaches to pro-active actions. Indeed, a frustrating aspect of digital transformation is its accelerated change. Ideas and concepts borrowed from communities-of-practice can help businesses deal with the evolutionary aspects of consumer uncertainty.

The relationship then between communities of practice and analysis and design is significant, given that future IT applications will heavily rely on informal inputs. While there may be attempts to computerize knowledge using predictive analytics software and big data, it will not be able to provide all of the risk associated behaviors of users and consumers. That is, a "structured" approach to creating predictive behavior reporting is typically difficult to establish and maintain.

Ultimately the dynamism from digital transformations creates too many uncertainties to be handled by sophisticated automated applications on how organizations will react to digital change variables. So, COP along with these predictive analytics applications provides a more thorough umbrella of how to deal with the ongoing and unpredictable interactions established by emerging digital technologies.

Support for the above position is found in the fact that technology requires accumulative collective learning that needs to be tied to social practices; this way, project plans can be based on learning as a participatory act. One of the major advantages of communities-of-practice is that it can integrate key competencies into the very fabric of the organization (Lesser et al. 2000). IT's typical disadvantage is that its staff needs to serve multiple organizational structures simultaneously. This requires that priorities be set by the organization. Unfortunately, it is difficult if not impossible for IT departments to establish such priorities without engaging in communities-of-practice concepts that allow for a more integrated process of negotiation and determination. Much of the process of COP would be initiated by digital disruptions and result in many behavioral changes—that is, the process of implementing communities-of-practice will necessitate changes in the analysis and design approach and fundamental organization processes as outlined in this book.

As stated above, communities-of-practice activities can be very significant in the analysis and design of transitioning requirements from a transitioning digital culture. According to Lesser et al. (2000), a knowledge strategy based on communities-of-practice consists of seven basic steps as shown in Table 2.4.

Lesser et al. (2000) suggest that communities of practice are heavily reliant on innovation. "Some strategies rely more on innovation than others for their success... once dependence on innovation needs have been clarified, you can work to create new knowledge where innovation matters" (8). Indeed, electronic communities of practice are different than physical communities. Digital disruptions provide another dimension to how technology affects organizational learning. It does so by creating new ways in which communities of practice operate. In the complexity of ways that it affects us, technology has a dichotomous relationship with communities of practice. That is, there is a two-sided issue: (1) the need for communities of practice to implement IT projects and integrate them better into what consumers want, and (2) the expansion of electronic communities of practice invoked by technology, which can in turn assist expansion of the business consumer base, globally and culturally.

The latter issue establishes the fact that a consumer can now readily be a member of many electronic communities and in many different capacities. Electronic communities are different in that they can have memberships that are short-lived and transient, forming and reforming according to interest, particular tasks, or commonality of issue. Communities of practice themselves are utilizing technologies to form multiple and simultaneous relationships. Furthermore, the growth of international communities resulting from ever-expanding global economies has created further complexities and dilemmas.

Table 2.4 Extended seven steps of community of practice strategy in analysis and design

Step	Communities-of-practice step	Analysis extension
1	Understanding strategic knowledge needs: what knowledge is critical to success	Understanding how technology affects strategic knowledge and what specific technological knowledge is critical to success
2	Engaging practice domains: where people form communities of practice to engage in and identify with	Technology identifies groups based on business-related benefits. Requiring domains to work together towards measurable results
3	Developing communities: how to help key communities reach their full potential	Technologies have life cycles that require communities to continue. Treating the life cycle as a supporter for attaining maturation and full potential
4	Working the boundaries: how to link communities to form broader learning systems	Technology life cycles require new boundaries to be formed. This will link other communities that were previously outside of discussions and thus expands input into technology innovations
5	Fostering a sense of belonging: how to engage people's identities and sense of belonging	The process of integrating communities: IT and other organizational units will create new evolving cultures which foster belonging as well as new social identities
6	Running the business: how to integrate communities of practice into running the business of the organization	Digital transformation provides new organizational structures that are necessary to operate communities of practice and to support new technological innovations
7	Applying, Assessing, Reflecting, Renewing: how to deploy knowledge strategy through waves of organizational transformation	The active process of dealing with multiple new technologies that accelerates the deployment of knowledge strategy. Emerging technologies increase the need for organizational transformation

Thus far I have presented communities of practice as an infrastructure that can foster improved ways to create requirements based on consumer behavior and trends. Most of what I have presented impacts the ways analysis and design needs to be approached in today's changing world. Communities of practice through the advent technology innovations have expanded to include electronic communities. While technology can provide organizations with vast electronic libraries that end up as storehouses of information, they are only valuable if it they are allowed to be shared within the community. Although IT has led many companies to imagine a new world of leveraged knowledge, communities have discovered that just storing information does not provide for effective and efficient use of knowledge. As a result, many companies have created these "electronic" communities so that knowledge can be leveraged, especially across cultures and geographic boundaries.

These electronic communities are predictably more dynamic as a result of what technology provides to them. Below are examples of what these communities provide to organizations.

- Transcending boundaries and exchanging knowledge with internal and external communities. In this circumstance, communities are not only extending across business units, but into communities among various clients—as we see developing in advanced e-business strategies. Using Internet and Intranets, communities can foster dynamic integration of the client, an important participant in competitive advantage. However, the expansion of an external community, due to emergent electronics, creates yet another need for the implementation of a more dynamic analysis approach to ascertaining requirements.
- Connecting social and workplace communities through sophisticated networks. This issue links well to the entire expansion of issues surrounding organizational learning, in particular learning organization formation. It enfolds both the process and the social dialectic issues so important to creating well-balanced communities of practice that deal with organizational-level and individual development.
- Integrating teleworkers and non-teleworkers, including the study of gender and cultural differences. The growth of distance workers will most likely increase with the maturation of technological connectivity. Video conferencing and improved media interaction through expanded broadband will support further developments in virtual workplaces. Gender and culture will continue to become important issues in the expansion of existing models that are currently limited to specific types of workplace issues.
- Assisting in computer-mediated communities. Such mediation allows for the management of interaction among communities, of who mediates their communications criteria, and of who is ultimately responsible for the mediation of issues. Mature communities of practice will pursue self-mediation.
- Creating "flame" communities. A "flame" is defined as a lengthy, often personally insulting, debate in an electronic community, which provides both positive and negative consequences. Difference can be linked to strengthening the identification of common values within a community, but requires organizational maturation that relies more on computerized communication to improve interpersonal and social factors to avoid miscommunications (Franco et al. 2000).
- Storing collective requirements in large-scale libraries and databases. As Einstein stated: "Knowledge is experience. Everything else is just information." Repositories of information are not knowledge, and they often inhibit organizations from sharing important knowledge building-blocks that affect technical, social, managerial, and personal developments that are critical for learning organizations (McDermott 2000).

Ultimately the above communities of practice are forming new social networks, which have established the cornerstone of "global connectivity, virtual communities, and computer-supported cooperative work" (Wellman et al. 2000, p. 179). These social networks are then creating new sources of trends, changing the very nature of the way organizations deal with and use technology to change how knowledge develops and is used via communities of practice. It is not, therefore, that communities of practice are new infrastructures or social forces, but rather the difference is in the way they communicate. Digital transformation forces new networks of communication to occur and the cultural adaptation component allows these communities of practice to focus on how they will use new emerging technologies to change their business and social lives.

2.8.1 Model-Driven AI

Model-driven AI captures knowledge and drives decisions via real representations and rules. In a model-driven AI, rules govern the definition of things and how they relate to others. The rules engine consists of defining relationship of the data and its related transaction capabilities. If an animal does not walk on two legs, then it has specific restrictions for those that use four legs for example. Most model driven AI is restricted where decision trees can determine specific rules that lead to certain paths of process. A perfect example of this limited type of AI was originally known as "Expert Systems" in which based on rules specific application paths were determined. A product that fits Expert Systems is tax returns, which are dictated by decision trees off of rule-based decisions, shown below in Fig. 2.7.

Once the individual makes a decision on status, the rules change based on the answer. Obviously one can see that the decision trees can become very complex, and may change. Indeed, tax programs have changes every year—so depending on the year selected, the product would perform differently. So the data (s versus m in

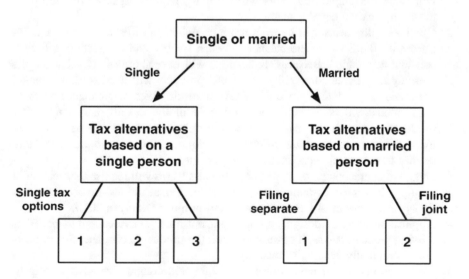

Fig. 2.7 Expert system rule-based decisions

this case) dictates the execution of the actual programs to be executed each time. If there are changes to the rules, per se, like a new tax law, then by changing the rule file, one can avoid making coding change.

Rule based inference engine products are part of the AI family, but they do not represent all of them. Those that use mathematical models and regression are beyond the scope of the analyst.

2.9 The Analyst in the Digital Transformation Era

When we discuss the digital world and its multitude of effects on how business is conducted and how the social world interacts, one must ask how this impacts the profession of analysis and design. This section attempts to address the perceived evolution of the role.

1. The analyst must become more innovative. While the business has the problem of keeping up with changes in their markets, the analyst needs to provide more solutions. Many of these solutions will not be absolute and likely will have short shelf lives. Risk is fundamental. As a result, analysts must truly become "business" analysts by exploring new ideas from the outside and continually considering how to implement the needs of the company's consumers. As a result, the business analyst will emerge as an idea broker (Robertson and Robertson 2012) by constantly pursuing external ideas and transforming them into automated and competitive solutions. These ideas will have a failure rate, which means that companies will need to produce more applications than they will inevitable implement. This will certainly require organizations to spend more on software development.

2. Quality requirements will be even more complex. In order to keep in equilibrium with the S-Curve the balance between quality and production will be a constant negotiation. Because applications will have shorter life cycles and the pressure to provide competitive solutions, products will need to sense market need and respond to them quicker. As a result, fixes and enhancements to applications will be become more inherent in the development cycle after products go live in the market. Thus, the object paradigm will become even more fundamental to better software development because it provides more readily tested reusable applications and routines.

3. Dynamic interaction among users and business teams will require the creation of multiple layers of communities of practice. Organizations involved in this dynamic process must have autonomy and purpose (Narayan 2015).

4. Application analysis, design, and development must be treated and managed as a living process, that is, it never ends until the product is obsolete. So products must continually develop to maturity.

5. Organizations should never outsource a new and competitive-advantage technology until it reaches commoditization.

2.10 Problems and Exercises

1. What is the relationship between digital transformation and analysis and design?
2. What are the benefits of obtaining an organization chart prior to conducting interviews with users?
3. How does politics affect the role of the analyst and his/her approach to the information-gathering function of the interviews?
4. Why does understanding user skills provide the analyst with an advantage during interviews
5. What are the six (6) sources of consumer analysis?
6. What is meant by requirements without users and without input?
7. Describe the relationship between technology and changes in the market. How does innovation play a pivotal role?
8. Describe Porter's Five Forces Framework.
9. Compare Porter's Five Forces with Langer's Sources of Consumer Analysis.
10. What is the S-Curve?
11. Explain the effect of digital transformation on the S-Curve.
12. What is an extended S-Curve?
13. How has risk factors related to digital transformation?
14. How has IoT and wireless affected the internet era?
15. What is a mini-loop in the context of a technology driver?
16. Explain the organization of a community of practice? Why is it important to the analyst in the digital age?
17. What is an Expert System? Explain.
18. What is meant by model-driven AI?

References

Elkjaer, B. (1999). In search of a social learning theory. In M. Easterby-Smith, J. Burgoyne, & L. Araujo (Eds.), *Organizational learning and the learning organization.* London: Sage.

Franco, V., Hu, H., Lewenstein, B. V., Piirto, R., Underwood, R., & Vidal, N. K. (2000). Anatomy of a flame: Conflict and community building on the Internet. In E. L. Lesser, M. A. Fontaine, & J. A. Slusher (Eds.), *Knowledge and communities* (pp. 209–224). Woburn, MA: Butterworth-Heinemann.

Langer, A. M. (1997). *The art of analysis.* New York: Springer.

Langer, A. M. (2011a). *Information technology and organizational learning: Managing behavioral change through technology and education* (2nd ed.). Boca Raton, FL: Taylor & Francis.

Langer, A. M. (2011b). *Information technology and organizational learning: Managing behavioral change through technology and education* (2nd ed.). New York: CRC Press.

Lesser, E. L., Fontaine, M. A., & Slusher, J. A. (Eds.). (2000). *Knowledge and communities.* Woburn, MA: Butterworth-Heinemann.

McDermott, R. (2000). Why information technology inspired but cannot deliver knowledge management. In E. L. Lesser, M. A. Fontaine, & J. A. Slusher (Eds.), *Knowledge and communities* (pp. 21–36). Woburn, MA: Butterworth-Heinemann.

Mintzberg, H., & Waters, J. A. (1985). Of strategies, deliberate and emergent. *Strategic Management Journal, 6,* 257–272.

Narayan, S. (2015). *Agile IT organization design for digital transformation and continuous delivery.* New York: Addison-Wesely.

Robertson, S., & Robertson, J. (2012). *Mastering the requirements process: Getting requirements right* (3rd ed.). Upper Saddle River, NJ: Addison-Wesley.

Wellman, B., Salaff, J., Dimitrova, D., Garton, L., Gulia, M., & Haythornthwaite, C. (2000). Computer networks and social networks: Collaborative work, telework, and virtual community. In E. L. Lesser, M. A. Fontaine, & J. A. Slusher (Eds.), *Knowledge and communities* (pp. 179–208). Woburn, MA: Butterworth-Heinemann.

Wenger, E. (1998). *Communities of practice: Learning, meaning and identity.* Cambridge, MA: Cambridge University Press.

Reviewing the Object Paradigm

3

This chapter will provide the historical structured analysis and design methodology that led to the object paradigm. At the core of an evolutionary approach are a set of traditional tools that need to be extended to meet the needs of an agile architecture in a mobile IoT market.

3.1 The Concept of the Logical Equivalent

The primary mission of an analyst or systems designer is to extract the physical requirements of the users and convert them to software. All software can trace its roots to a physical act or a physical requirement. A physical act can be defined as something that occurs in the interaction of people, that is, people create the root requirements of most systems, especially those in business. For example, when Mary tells us that she receives invoices from vendors and pays them thirty days later, she is explaining her physical activities during the process of receiving and paying invoices. When the analyst creates a technical specification, which represent Mary's physical requirements, the specification is designed to allow for the translation of her physical needs into an automated environment. We know that software must operate within the confines of a computer, and such systems must function on the basis of logic. The logical solution does not always treat the process using the same procedures employed in the physical world. In other words, the software system implemented to provide the functions which Mary does physically will probably work differently and more efficiently than Mary herself. Software, therefore, can be thought of as a logical equivalent of the physical world. This abstraction, which I call the concept of the Logical Equivalent (LE), is a process that analysts must use to create effective requirements of the needs of a system. The LE can be compared to a schematic of a plan or a diagram of how a technical device works.

© Springer Nature Switzerland AG 2020
A. M. Langer, *Analysis and Design of Next-Generation Software Architectures*,
https://doi.org/10.1007/978-3-030-36899-9_3

Any success in creating a concise and accurate schematic of software that needs to be developed by a programmer will be directly proportional to how well the analyst masters Langer's (1997) Concept of the Logical Equivalent. Very often requirements are developed by analysts using various methods that do not always contain a basis for consistency, reconciliation and maintenance. There is usually far too much prose used as opposed to specific diagramming standards that are employed by engineers. After all, we are engineering a system through the development of software applications. The most critical step in obtaining the LE is the understanding of the process of Functional Decomposition. Functional Decomposition is the process for finding the most basic parts of a system, like defining all the parts of a car so that it can be built. It would be possible not from looking at a picture of the car, but rather at a schematic of all the functionally decomposed parts. Developing and engineering software is no different and essential to create reusable component applications that operate in the IoT environment.

Below is an example of an analogous process using functional decomposition, with its application to the LE.

In obtaining the physical information from the user, there are a number of modeling tools that can be used. Each tool provides a specific function to derive the LE. The word "derive" has special meaning here. It relates to the process of Long Division, or the process or formula we apply when dividing one number by another. Consider the following example:

```
         256  remainder 4    } Result or Answer
    5 | 1284                  } Problem to Solve
        10
        284
        25                    } Formula applied to produce result or answer
        34
        30
         4
```

The above example shows the formula that is applied to a division problem. We call this formula long division. It provides the answer, and if we change any portion of the problem, we simply re-apply the formula and generate a new result. Most important, once we have obtained the answer, the value of the formula steps is only one of documentation. That is, if someone questioned the validity of the result, we could show them the formula to prove that the answer was correct (based on the input).

Now let us apply long division to obtaining the LE via functional decomposition. The following is a result of an interview with Joe, a bookkeeper, about his physical procedure for handling bounced checks.

> Joe the bookkeeper receives bounced checks from the bank. He fills out a Balance Correction Form and forwards it to the Correction Department so that the outstanding balance can be corrected. Joe sends a bounced check letter to the customer requesting a replacement check plus a $15.00 penalty (this is now included as part of the outstanding balance). Bounced checks are never re-deposited.

The appropriate modeling tool to use in this situation is a Data Flow Diagram (DFD). A DFD is a tool that shows how data enters and leaves a particular process. The process we are looking at with Joe is the handling of the bounced check. A DFD has four possible components:

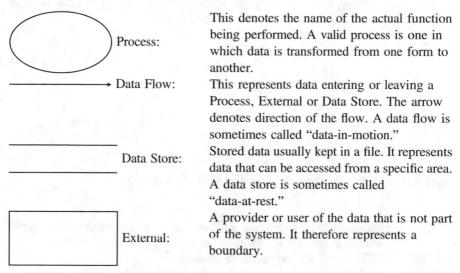

Process: This denotes the name of the actual function being performed. A valid process is one in which data is transformed from one form to another.

Data Flow: This represents data entering or leaving a Process, External or Data Store. The arrow denotes direction of the flow. A data flow is sometimes called "data-in-motion."

Data Store: Stored data usually kept in a file. It represents data that can be accessed from a specific area. A data store is sometimes called "data-at-rest."

External: A provider or user of the data that is not part of the system. It therefore represents a boundary.

Now let us draw the LE of Joe's procedure using DFD tools as shown in Fig. 3.1.

The above DFD shows that bounced checks arrive from the bank, the Account Master file is updated, the Correction Department is informed and Customers receive a letter. The Bank, Correction Department and Customers are considered "outside" the system and are therefore represented logically as Externals. This

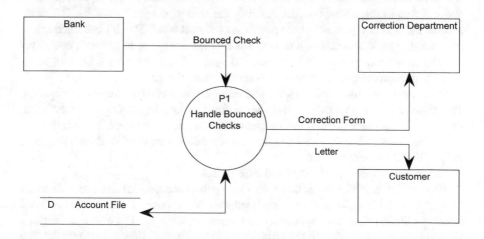

Fig. 3.1 Data flow diagram for handling bounced checks

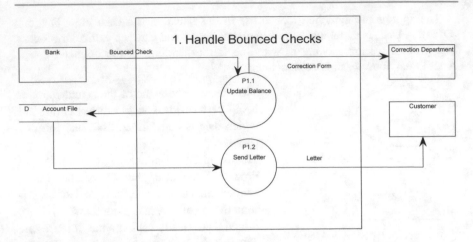

Fig. 3.2 Level 2 data flow diagram for handling bounced checks

diagram is considered to be at the first level or "Level 1" of functional decomposition. You will find that all modeling tools employ a method to functionally
decompose. DFDs use a method called "Leveling."

The question is whether we have reached the most basic parts of this process or
should we level further. Many analysts suggest that a fully decomposed DFD
should have only one data flow input and one data flow output. Our diagram
currently has many inputs and outputs and therefore it can be leveled further. The
result of functionally decomposing to the second level (Level 2) is as shown in
Fig. 3.2.

Notice that the functional decomposition shows us that Process 1: Handling
Bounced Checks is really made up of two sub-processes called 1.1 Update Balance
and 1.2 Send Letter. The box surrounding the two processes within the Externals
reflects them as components of the previous or parent level. The double-sided arrow
in Level 1 is now broken down to two separate arrows going in different directions
because it is used to connect Processes 1.1 and 1.2. The new level is more functionally decomposed and a better representation of the LE.

Once again, we must ask ourselves whether Level 2 can be further decomposed.
The answer is yes. Process 1.1 has two outputs to one input. On the other hand,
Process 1.2 has one input and one output and is therefore complete. 1.2 is said to be
at the Functional Primitive, a DFD that cannot be decomposed further. Therefore,
only 1.1 will be decomposed.

Let us decompose 1.1 as depicted in Fig. 3.3.

Process 1.1 is now broken down into two sub processes: 1.1.1 Update Account
Master and 1.1.2 Inform Correction Department. Process 1.1.2 is a Functional
Primitive since it has one input and one output. Process 1.1.1 is also considered a
Functional Primitive because the "Bounced Check Packet" flow is between the two

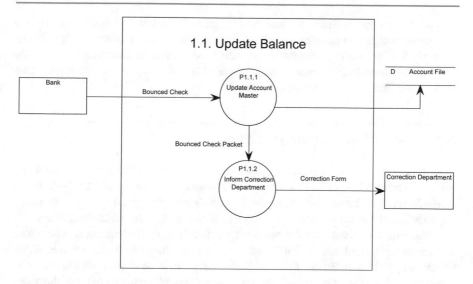

Fig. 3.3 Level 3 data flow diagram for handling bounced checks

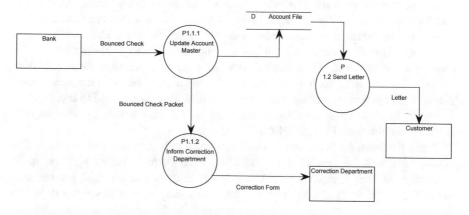

Fig. 3.4 Functionally decomposed level 3 data flow diagram for handling bounced checks

processes and is used to show connectivity only. functional decomposition is at Level-3 and is now complete.

The result of functional decomposition is the following DFD (Fig. 3.4).

As in long division, only the complete result, represented above, is used as the answer. The preceding steps are formulas that we use to get to the lowest, simplest representation of the logical equivalent. Levels 1, 2 and 3 are used only for documentation of how the final DFD was determined.

The logical equivalent is an excellent method that allows analysts and systems designers to organize information obtained from users and to systematically derive the most fundamental representation of their process. It also alleviates unnecessary pressure to immediately understand the detailed flows and provides documentation of how the final schematic was developed.

3.2 Tools of Structured Analysis

Now that we have established the importance and goals of the logical equivalent, we can turn to a discussion of the methods available to assist the analyst. These methods serve as the tools to create the best models in any given situation, and thus the most exact logical equivalent. The tools of the analyst are something like those of a surgeon, who uses only the most appropriate instruments during an operation. It is important to understand that the surgeon is sometimes faced with choices about which surgical instruments to use; particularly with new procedures, there is sometimes disagreement among surgeons about which instruments are the most effective. The choice of tools for analysis and data processing is no different; indeed, it can vary more and be more confusing. The medical profession, like many others, is governed by its own ruling bodies. The American Medical Association and the American College of Physicians and Surgeons, as well as state and federal regulators, represent a source of standards for surgeons. Such a controlling body does not exist in the data processing industry, nor does it appear likely that one will arise in the near future. Thus, the industry has tried to standardize among its own leaders. The result of such efforts has usually been that the most dominant companies and organizations create standards to which others are forced to comply. For example, Microsoft has established itself as an industry leader by virtue of its software domination. Here, Might is Right!

Since there are no real formal standards in the industry, the analysis tools discussed here will be presented on the basis of both their advantages and their shortcomings. It is important then to recognize that no analysis tool (or methodology for that matter) can do the entire job, nor is any perfect at what it does. To determine the appropriate tool, analysts must fully understand the environment, the technical expertise of users and the time constraints imposed on the project. By "environment" we mean the existing system and technology, computer operations, and the logistics–both technically and geographically–of the new system. The treatment of the user interface should remain consistent with the guidelines discussed in Chap. 2.

The problem of time constraints is perhaps the most critical of all. The tools you would ideally like to apply to a project may not fit the time frame allotted. What happens, then, if there is not enough time? The analyst is now faced with selecting a second-choice tool that undoubtedly will not be as effective as the first one would have been. There is also the question of how tools are implemented, that is, can a hybrid of a tool be used when time constraints prevent full implementation of the desired tool?

3.3 Making Changes and Modifications

Within the subject of analysis tools is the component of maintenance modeling, or how to apply modeling tools when making changes or enhancements to an existing product. Maintenance modeling falls into two categories:

1. Pre-modeled: where the existing system already has models that can be used to effect the new changes to the software.
2. Legacy System: where the existing system has never been modeled; any new modeling will therefore be incorporating analysis tools for the first time.

Pre-modeled:

Simply put, a Pre-Modeled product is already in a structured format. A structured format is one that employs a specific format and methodology such as the data flow diagram.

The most challenging aspects of changing Pre-Modeled tools are:

1. keeping them consistent with their prior versions, and
2. implementing a version control system that provides an audit-trail of the analysis changes and how they differ from the previous versions. Many professionals in the industry call this Version Control; however, care should be taken in specifying whether the version control is used for the maintenance of analysis tools. Unfortunately, Version Control can be used in other contexts, most notably in the tracking of program versions and software documentation. For these cases, special products exist in the market which provide special automated "version control" features. We are not concerned here with these products but rather with the procedures and processes that allow us to incorporate changes without losing the prior analysis documentation. This kind of procedure can be considered consistent with the long division example in which each time the values change, we simply re-apply the formula (methodology) to calculate the new answer. Analysis version control must therefore have the ability to take the modifications made to the software and integrate them with all the existing models as necessary.

Being Consistent:

It is difficult to change modeling methods and/or CASE tools in the middle of the life cycle of a software product. One of our main objectives then is to try avoid doing so. How? Of course, the simple answer is to select the right tools and CASE software the first time. However, we all make mistakes, and more importantly, there are new developments in systems architecture that may make a new CASE product attractive. You would be wise to foresee this possibility and prepare for inconsistent tools implementation. The best offense here is to:

- ensure that your CASE product has the ability to transport models through an ASCII file or cut/paste method. Many have interfaces via an "export" function. Here, at least, the analyst can possibly convert the diagrams and data elements to another product.
- keep a set of diagrams and elements that can be used to establish a link going forward, that is, a set of manual information that can be re-input to another tool. This may be accomplished by simply having printed documentation of the diagrams; however, experience has shown that it is difficult to keep such information up to date. Therefore, the analyst should ensure that there is a procedure for printing the most current diagrams and data elements.

Should the organization decide to use different tools, e.g., process-dependency diagrams instead of data flow diagrams, or a different methodology such as crows-foot method in Entity Relational Diagramming, then the analyst must implement a certain amount of re-engineering. This means mapping the new modeling tools to the existing ones to ensure consistency and accuracy. This is no easy task, and it is strongly suggested that you document the diagrams so you can reconcile them.

Version Control:

This book is not intended to focus on the generic aspects of version control; however, structured methods must have audit trail. When a new process is changed, a directory should be created for the previous version. The directory name typically consists of the version and date such as: xyz1.21295, where xyz is the name of the product or program, 1.2 the version and 1295 the version date. In this way previous versions can be easily re-created or viewed. Of course, saving a complete set of each version may not be feasible or may be too expensive (in terms of disk space, etc.). In these situations, it is advisable to back up the previous version in such a manner as to allow for easy restoration. In any case, a process must exist, and it is crucial that there be a procedure to do backups periodically.

3.4   What is Object-Oriented Analysis?

Object-Oriented Analysis is the key analysis tool in the design of successful mobile applications. It is without question the most important element of creating what may be called the "complete" requirement agile system. There are a number of approaches used by the industry and perhaps controversy about the best approach and tools that should be used to create mobile-object systems. This chapter will focus on developing the requirements for object systems and the challenges of converting legacy systems. Therefore, many of the terms will be defined based on their fundamental capabilities and how they can be used by a practicing analyst (as opposed to a theorist!).

Fig. 3.5 A car is an example of a physical object

Object Orientation (OO) is based on the concept that every requirement ulti-mately must belong to an object. It is therefore critical that we first define what is meant by an object. In the context of OO analysis, an object is any cohesive whole made up of two essential components: data and processes (Fig. 3.5).

Traditional analysis approaches were traditionally based on the examination of a series of events. We translated these events from the physical world by first inter-viewing users and then developing what was introduced as the concept of the logical equivalent. Although we are by no means abandoning this necessity, the OO paradigm requires that these events belong to an identifiable object. Let us expand on this difference using the object shown below, an object we commonly call a "car."

The above car may represent a certain make and model, but it also contains common components that are contained in all cars (e.g., an engine). If we were to look upon the car as a business entity of an organization, we might find that the following three systems were developed over the years.

The above diagram shows us that the three systems were built over a period of 21 years. Each system was designed to provide service to a group of users responsible for particular tasks. The diagram shows that the requirements for System 1 were based on the engine and front-end of the car. The users for this project had no interest in or need for any other portions of the car. System 2, on the other hand, focused on the lower center and rear of the car. Notice, however, that System 2 and System 1 have an overlap. This means that there are parts and procedures common to both systems. Finally, System 3 reflects the upper center and rear of the car and has an overlap with System 2. It is also important to note that there are components of the car that have not yet been defined, probably because no user has had a need for them. We can look at the car as an object and Systems 1–3 as the software which has so far been defined about that object. Our observations should also tell us that the entire object is not defined and more important, that there is probable overlap of data and functionality among the systems that have been developed. This case exemplifies the history of most development systems. It should be clear that the users who stated their requirements never had any under-standing that their own situation belonged to a larger composite object. Internal users tend to establish requirements based on their own job functions and their own experiences in those functions. Therefore, the analyst who interviews users about their events is exposed to a number of risks:

- Users tend to identify only what they have experienced, rather than speculating about other events that could occur. This is a significant limitation in the mobile world and in attempting to understand what consumer *may* want in the future. We know that such events can take place, although they have not yet occurred (you should recall the discussion of using STDs as a modeling tool to identify unforeseen possibilities). Consider, for example, an analysis situation in which $50,000 must be approved by the firm's Controller. This event might show only the approval, not the rejection. The user's response is that the Controller, while examining the invoices, has never rejected one and therefore no rejection procedure exists. You might ask why. Well, in this case the Controller was not reviewing the invoices for rejection but rather holding them until he/she was confident that the company's cash flow could support the issuance of these invoices. Obviously, the Controller could decide to reject an invoice. In such a case, the software would require a change to accommodate this new procedure. From a software perspective we call this a system enhancement, and it would result in a modification to the existing system.
- Other parts of the company may be affected by the Controller's review of the invoices. Furthermore, are we sure that no one else has automated this process before? One might think such prior automation could never be overlooked, especially in a small company, but when users have different names for the same thing (remember Customer and Client!) it is very likely that such things will occur. In this example there were two situations where different systems overlapped in functionality.
- There will be conflicts between the systems with respect to differences in data and process definitions. Worst of all, these discrepancies may not be discovered until years after the system is delivered.

The above example shows us that requirements obtained based on an individual's events require another level of reconciliation to ensure they are accurate. Requirements are said to be "complete" when they define the whole object. The more incomplete they are, the more modifications likely will be required later. The more modifications in a system, the higher the likelihood that data and processes across applications may conflict with each other. Ultimately this results in a less dependable, lower quality system. Most of all, event analysis alone is prone to missing events that users have never experienced. This situation is represented in the car example by the portions of the car not included in any of the three systems. System functions and components may also be missed because users are absent or unavailable at the time of the interviews, or because no one felt the need to automate a certain aspect of the object. In either case, the situation should be clear. We need to establish objects prior to doing event analysis.

Before we discuss the procedures for identifying an object, it is worth looking at the significant differences between the object approach and earlier approaches. The first major systems were developed in the 1960s and were called Batch, meaning that they typically operated on a transaction basis. Transactions were collected and then used to update a master file. Batch systems were very useful in the financial industries, including banks. We might remember having to wait until the morning after a banking transaction to see our account balance because a batch process updated the master account files overnight. These systems were built based on event interviewing, where programmer/analysts met with users and designed the system. Most of these business systems were developed and maintained using COBOL.

In the early seventies, the new buzz word was "on-line, real-time" meaning that many processes could now update data immediately or on a "real-time" basis. Although systems were modified to provide these services, it is important to understand that they were not re-engineered. That is, the existing systems, which were based on event interviews, were modified, but not redesigned.

In the late 80s and early 90s the hot term became "Client/Server." These systems, which will be discussed later, are based on sophisticated distributed systems concepts. Information and processes are distributed among many Local and Wide Area Networks. Many of these client/server systems are re-constructions of the on-line real-time systems which in turn were developed from the 1960s batch systems. The point here is that we have been applying new technology to systems that were designed over 30 years ago without considering the obsolescence of the design.

Through these three generations of systems, the analyst has essentially been on the outside looking in (see Fig. 3.6). The completeness of the analysis was dependent upon–and effectively dictated by–the way the inside users defined their business needs.

OO, on the other hand, requires that the analyst have a view from the inside looking out. What we mean here is that the analyst first needs to define the generic aspects of the object and then map the user views to the particular components that exist within the object itself (Fig. 3.7). The diagram below shows a conceptual view of the generic components that could be part of a bank.

Figure 3.8 shows the essential functions of the bank. The analyst is on the inside of the organization when interviewing users and therefore will have the ability to map a particular requirement to one or more of its essential functions. In this approach, any user requirement must fit into at least one of the essential components. If a user has a requirement that is not part of an essential component, then it must be either qualified as missing (and thus added as an essential component) or rejected as inappropriate.

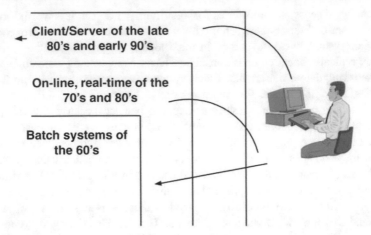

Fig. 3.6 Requirements are often developed by analysts from an outside view. The specifications are therefore dependent on the completeness of the user's view

The process of taking user requirements and placing each of their functions into the appropriate essential component can be called mapping. The importance of mapping is that functions of requirements are logically placed where they generically belong, rather than according to how they are physically implemented. For example, suppose Joseph, who works for a bank, needed to provide information to a customer about the bank's investment offerings. Joseph would need to access investment information from the system. If OO methods were used to design the system, all information about banking investments would be grouped together generically. Doing it this way allows authorized personnel to access investment information regardless of what they do in the bank. If event analysis alone was used, Joseph would probably have his own subsystem that defines his particular requirements for accessing investment information. The problem here is twofold: first, the subsystem does not contain all of the functions relating to investments. Should Joseph need additional information, he may need an enhancement or need to use someone else's system at the bank. Second, Joseph's subsystem may define functions that have already been defined elsewhere in another subsystem. The advantage of OO is that it centralizes all of the functions of an essential component and allows these functions to be "reused" by all processes that require its information. The computer industry calls this capability Reusable Objects.

3.5   Identifying Objects and Classes

The most important challenge of successfully implementing OO is the ability to understand and select Objects. We have already used an example which identified a car as an object. This example is what can be called the tangible object, or as the

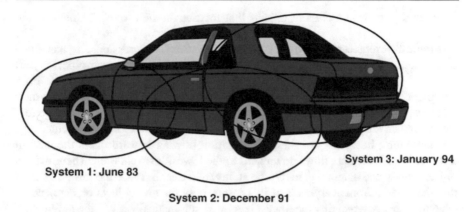

Fig. 3.7 This diagram reflects the three systems developed to support the car object

industry calls them "physical objects." Unfortunately, there is another type of object called an "abstract" or intangible object. An intangible object is one that you cannot touch or as Grady Booch originally described: "something that may be apprehended intellectually...Something towards which thought or action is directed."[1] An example of an intangible object is the security component of the essentials of the bank. In many instances OO analysis will begin with identifying tangible objects which will in turn make it easier to discover the intangible ones.

OO is somewhat consistent with the architecture of process and data in that all objects contain their own data and processes, called attributes and services, respectively. Attributes are effectively a list of data elements which are permanent components of the object. For example, a steering wheel is a data element that is a permanent attribute of the object "Car." The services (or operations), on the other hand, define all of the processes that are permanently part or "owned" by the object. "Starting the Car" is a service that is defined within the object car. This service contains the algorithms necessary to start a car. Services are defined and invoked through a method. A method is a process specification for an operation (service).[2] For example, "Driving the Car" could be a method for the car object. The Driving the Car" method would invoke a service called "Starting the Car" as well as other services until the entire method requirement is satisfied. Although a service and method can have a one-to-one relationship, it is more likely that a service will be a subset or be one of the operations that make up a method.

Objects have the ability to inherit attributes and methods from other objects when they are placed within the same class. A class is a group of objects that have similar attributes and methods and typically have been put together to perform a specific task. To further understand these concepts, we will establish the object for

---

[1]Booch (1995).
[2]Martin (1994).

"Car" and place it in a class of objects that focuses on the use of transmissions in cars.

Figure 3.9 represents an object class called Car Transmissions. It has three component objects: cars, automatic trans, and standard trans. The car object is said to be the parent object. Automatic trans and standard trans are object types. Both automatic trans and standard trans will inherit all attributes and services from their parent object cars. Inheritance in object technology means that the children effectively contain all of the capabilities of their parents. Inheritance is implemented as a tree structure[3]; however, instead of information flowing upward (as is the case in tree structures), the data flows downward to the lowest level children. Therefore, an object inheritance diagram is said to be an inverted tree. Because the lowest level of the tree inherits from everyone's of its parents, only the lowest level object need be executed, that is, executing the lowest level will automatically allow the application to inherit all of the parent information and applications as needed. We call the lowest level objects concrete, while all others in the class are called abstract. Objects within classes can change simply by the addition of a new object. Let us assume that there is another level added to our example. The new level contains objects for the specific types of automatic and standard transmissions.

The class in Fig. 3.10 has been modified to include a new concrete layer. Therefore, the automatic trans object and standard trans object are now abstract. The new four concrete objects not only inherit from their respective parent objects, but also from their common grandparent, cars. It is also important to recognize that classes can inherit from other classes. Therefore, the same example could show each object as a class: that is, cars would represent a class of car objects and automatic trans another class of objects. Therefore, the class automatic trans would inherit from the cars class in the same manner described above. We call this "class inheritance."

I mentioned before the capability of OO objects to be reusable (Re-usable Objects). This is very significant in that it allows a defined object to become part of another class, while still keeping its own original identity and independence. The example below demonstrates how Cars can be reused in another class (Fig. 3.11).

Notice that the object Car is now part of another class called Transportation Vehicles. However, Car, instead of being an abstract object within its class, has become concrete and thus inherits from its parent, Transportation Vehicles. The object Cars has methods that may execute differently depending on the class it is in. Therefore, Cars in the Transportation Vehicle class might interpret a request for "driving the car" as it relates to general transportation vehicles. Specifically, it might invoke a service that shows how to maneuver a car while it is moving. On the other hand, Cars in the Transmission class might interpret the same message coming from one of its children objects as meaning how the transmission shifts

---

[3]A data structure containing zero or more nodes that are linked together in a hierarchical fashion. The topmost node is called the root. The root can have zero or more child nodes, connected by links; the root is the parent node to its children. Each child node can in turn have zero or more children of its own. Microsoft Press, Computer Dictionary, Second Edition, p. 397.

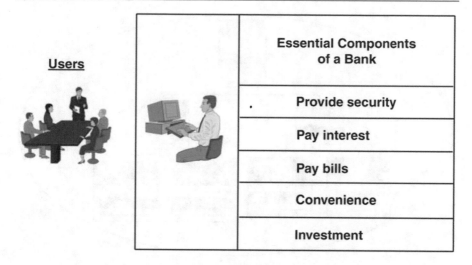

Fig. 3.8 Using the object approach, the analyst interviews users from the inside looking out

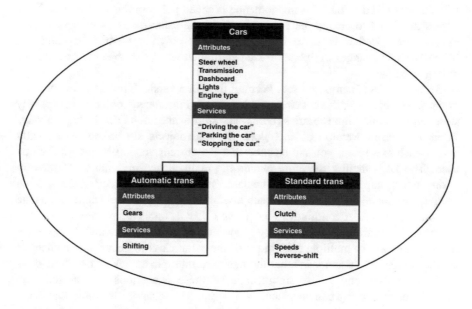

Fig. 3.9 Class car transmissions

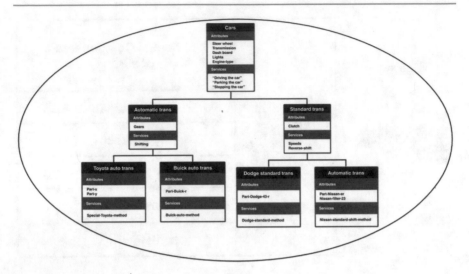

Fig. 3.10 Class car transmission types

when a person is driving. This phenomenon is called polymorphism. Polymorphism allows an object to change its behavior within the same methods under different circumstances. What is more important is that polymorphism is dynamic in behavior so its changes in operation are determined when the object is executed or during run-time.

Because objects can be reused, keeping the same version current in every copy of the same object in different classes is important. Fortunately, objects are typically stored in Dynamic Link Libraries (DLL). The significance of a DLL is that it always stores the current version of an object. Because objects are linked dynamically before each execution, you are ensured that the current version is always the one used. The DLL facility therefore avoids the maintenance nightmares of remembering which applications contain the same sub-programs. Legacy systems often need to re-link every copy of the subprogram in each module where a change occurs. This problem continues to haunt the COBOL application community.

Another important feature in object systems is Instantiation and Persistence. Instantiation allows multiple executions of the same class to occur independent of another execution. This means that the there are multiple copies of the same class executing concurrently. The significance of these executions is that they are mutually exclusive and can be executing different concrete objects within that class. Because of this capability, we say that objects can have multiple *instances* within each executing copy of a class it belongs to. Sometimes, although class executions are finished, a component object continues to operate or *persist*. Persistence is therefore an object that continues to operate after the class or operation that invoked it has finished. The system must keep track of each of these object instances.

The abilities of objects and classes to have inheritance, polymorphic behavior, instantiation and persistence are just some of the new mechanisms that developers can take advantage of when building OO systems.[4] Because of this, the analyst must not only understand the OO methodology, but must also apply new approaches and tools that will allow an appropriate schematic to be produced for system developers.

3.6 Object Modeling

Another analysis modeling tool is called a State Transition Diagram (STD) and useful for modeling event driven and time dependent systems. A state very closely resembles an object/class and therefore can be used with little modification to depict the flow and relationships of objects. The major difference between an object and a state is that an object is responsible for its own data (which we call an attribute in OO). An object's attributes are said to be *encapsulated* behind its methods, that is, a user cannot ask for data directly. The concept of encapsulation is that access to an object is allowed only for a purpose rather than for obtaining specific data elements. It is the responsibility of the method and its component services to determine the appropriate attributes that are required to service the request of the object. An object diagram, regardless of whose methodology is used, is essentially a hybrid of an STD and an Entity Relational Diagram (ERD). The STD represents the object's methods and the criteria for moving from one object to another. The ERD, on the other hand, defines the relationship of the attributes between the stored data models. The result is best shown using the order processing example below.

Figure 3.12 reflects that a customer object submits a purchase order for items to the order object. The relationship between customer and order reflects both STD and ERD characteristics. The "submits purchase order" specifies the condition to change the state of or move to the order object. The direction arrow also tells us that the order object cannot send a purchase order to the customer object. The crow's foot cardinality shows us that a customer object must have at least one order to create a relationship with the order object. After an order is processed, it is prepared for shipment. Notice that each order has one related shipment object; however multiple warehouse items can be part of a shipment. The objects depicted above can also represent classes suggesting that they are comprised of many component objects. These component objects might in turn be further decomposed into other

---

[4]This book is not intended to provide all of the specific technical capabilities and definitions that comprise the OO paradigm, but rather its effects on the analyst approach. Not all of the OO issues are analyst responsibilities, and many of them are product-specific. Because OO is still very controversial, OO products are not consistent in their use of OO facilities. For example, C+ + allows multiple inheritance meaning that a child can have many parent objects. This is inconsistent with the definition of a class as a tree structure since children in tree structures can have only one parent.

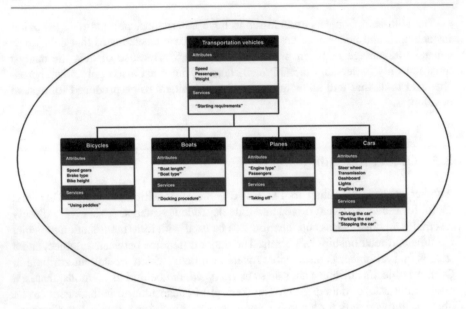

Fig. 3.11 Class transportation vehicles

primitive objects. This is consistent with the concept of the logical equivalent and with functional decomposition (Fig. 3.13).

It is important that the analyst specify whether classes or objects are depicted in the modeling diagrams. It is not advisable to mix classes and objects at the same level. Obviously, the class levels can be effective for user verification, but objects will be inevitably required for final analysis and engineering.

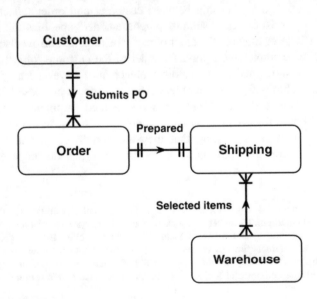

Fig. 3.12 An object/class diagram

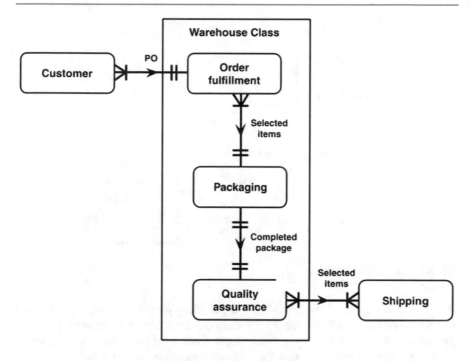

Fig. 3.13 The component objects of the warehouse class

3.7 Relationship to Structured Analysis

Many analysts make the assumption that the traditional structured tools are not required in OO analysis. This simply is not true, as we have shown in the previous examples. To further emphasize the need to continue using structured techniques, we need to understand the underlying benefit of the OO paradigm and how structured tools are necessary to map to the creation of objects and classes. It is easy to say: "find all the objects in the essential components"; actually, to have a process to do so is another story. Before providing an approach to determine objects, let us first understand the problem.

3.7.1 Application Coupling

Coupling can be defined as the measurement of an application's dependency on another. Simply put, does a change in an application program necessitate a change to another application program? Many known system malfunctions have resulted from highly coupled systems. The problem, as you might have anticipated, relates back to the analysis function, where decisions could be made as to what services

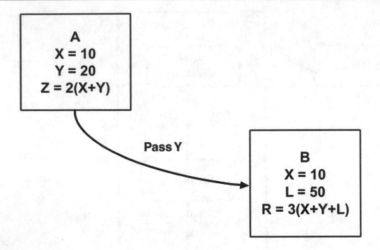

Fig. 3.14 Application coupling

should be joined to form one single application program. Coupling is never something that we want to do, but no system can be made up of just one program. Therefore, coupling is a reality and one that analysts must focus on. Let us elaborate on the coupling problem through the following example (Fig. 3.14).

The two programs A and B are coupled via the passing of the variable Y. Y is subsequently used in B to calculate R. Should the variable Y change in A, it will not necessitate a change in B. This is considered good coupling. However, let us now examine X. We see that X is defined in both A and B. Although the value of X does not cause a problem in the current versions of A and B, a subsequent change of X

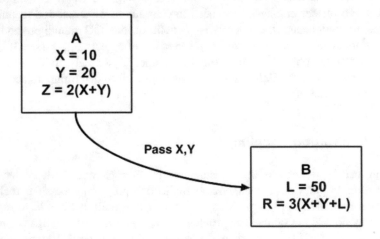

Fig. 3.15 Application coupling using variables X and Y

will cause a programmer to remember to change the value in B. This is a maintenance nightmare. In large enterprise level systems, analysts and programmers cannot "remember" where all of these couples have occurred, especially when the original developers are no longer with the organization. The solution to this problem is also to pass X from program A as shown in Fig. 3.15.

We now see that both X and Y are passed and programs A and B are said to have low coupling. In addition, program A is said to be more *cohesive*.

3.7.2 Application Cohesion

Cohesion is the measurement of how independent a program is on its own processing. That is, a cohesive program contains all of the necessary data and logic to complete its applications without being directly affected by another program; a change in another program should not require a change to a cohesive one. Furthermore, a cohesive program should not cause a change to be made in another program. Therefore, cohesive programs are independent programs that react to messages to determine what they need to do; however, they remain self-contained. When program A also passed X it became more cohesive because a change in X no longer required a change to be made to another program. In addition, B is more cohesive because it gets the change of X automatically from A. Systems that are designed more cohesively are said to be more maintainable. Their codes can also be reused or retrofitted into other applications as components because they are wholly independent. A cohesive program can be compared to an interchangeable standard part of a car. For example, if a car requires a standard 14-in. tire, typically any tire that meets the specification can be used. The tire, therefore, is not married to the particular car, but rather is a cohesive component for many cars.

Cohesion is in many ways is the opposite of coupling. The higher the cohesion, the lower the coupling. Analysts must understand that an extreme of either cohesion or coupling cannot exist. This is shown in the graph below (Fig. 3.16).

The graph shows that we can never reach 100% cohesion; that would mean there is only one program in the entire system, a situation that is unlikely. However, it is possible to have a system where a 75% cohesion ratio is obtained.

We now need to relate this discussion to OO. Obviously OO is based very much on the concept of cohesion. Objects are independent reusable modules that control their own attributes and services. Object coupling is based entirely on message processing via inheritance or collaboration.[5] Therefore, once an object is identified, the analyst must define all of its processes in a cohesive manner. Once the cohesive processes are defined, the required attributes of the object are then added to the object. Below is a table which shows how processes can be combined to create the best cohesion (Fig. 3.17).

---

[5]Collaboration is the interaction between objects and classes where inheritance is not used. Inheritance can operate only in hierarchical structures; however, many object and class configurations can simply "talk" to one another through messaging systems.

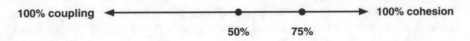

Fig. 3.16 Coupling and cohesion relationships

The tiers above are based on best to worst, where By Function is the most desirable and By Lines of Code the least desirable. Tiers 1 and 2 will render the best object cohesiveness. This can be seen with the following example.

Figure 3.18 depicts a four-screen system that includes four objects, that is, each screen is a separate object. The Transaction Processing object has been designed using Tier 2, By Same Data since it deals only with the Transaction File. The object is cohesive because it does not depend on or affect another module in its processing. It provides all of the methods required for transaction data.

The Financials object is an example of Tier 1, By Function since a Balance Sheet is dependent on the Income Statement and the Income Statement is dependent on the Trial Balance. The object therefore is self-contained within all the functions necessary to produce financial information (in this example).

The System Editor, on the other hand, being an example of Tier 3, shows that it handles all of the editing (verification of the quality of data) for the system. Although there appears to be some benefit to having similar code in one object, we can see that it affects many different components. It is therefore considered a highly coupled object and not necessarily the easiest to maintain.

We can conclude that Tiers 1 and 2 provide analysts with the most attractive way for determining an object's attributes and services. Tiers 3 and 4, although practiced, do not provide any real benefits in OO and should be avoided as much as

Tier	Method	Method Description
1	By Function	Processes are combined into one object / class based on being a component of the same function. Examples include: Accounts Receivable, Sales, and Goods Returned are all part of the same function. A sale creates a receivable and goods returned decreases the sale and the receivable.
2	By Data	Processes are combined based on their use of the same data and data files. Processes that tend to use the same data are more cohesive.
3	By Generic Operation	Processes are combined based on their generic performance. Examples could be "editing" or "printing."
4	By Lines of Code	Processes are created after an existing one reaches a maximum number of lines in the actual program source code.

Fig. 3.17 Methods of selecting cohesive objects

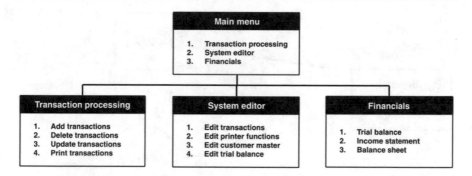

Fig. 3.18 Applications with varying types of object cohesion

possible. The question now is what technique do we follow to start providing the services and attributes necessary when developing logical objects?

The structured tools discussed in Chap. 3 provide us with the essential capabilities to work with OO analysis and design. The STD can be used to determine the initial objects and the conditions of how one object couples or relates to another. Once the STD is prepared it can be matured into the object model discussed earlier in this chapter. The object model can be decomposed to its lowest level; the attributes and services of each object must then be defined. All of the DFD functional primitives can now be mapped to their respective objects as services within their methods. It is also a way of determining whether an object is missing (should there be a DFD that does not have a related object). The analyst should try to combine each DFD using the Tier 1 By Function approach. This can sometimes be very difficult depending on the size of the system. If the Tier 1 approach is too difficult, the analyst should try Tier 2 by combining DFDs based on their similar data stores. This is a very effective approach; since Tier 1 implies Tier 2,[6] it is a very productive way to determine how processes should be mapped to their appropriate objects. This does not suggest that the analyst should not try Tier 1 first.

The next activity is to determine the object's attributes or data elements. The ERD serves as the link between an attribute in an object and its actual storage in a database. It is important to note that the attribute setting in an object may have no resemblance to its setting in the logical and physical data entity. The data entity is focused on the efficient storage of the elements and its integrity, whereas the attribute data in an object is based on its cohesiveness with the object's services.

The mapping of the object to the DFD and ERD can be best shown graphically below (Fig. 3.19).

---

[6]We have found that application programs that have been determined using Tier 1 will always imply Tier 2. That is, applications that are combined based on function typically use the same data. Although the converse is not necessarily true, we believe it is an excellent approach to backing-into the functions when they are not intuitively obvious.

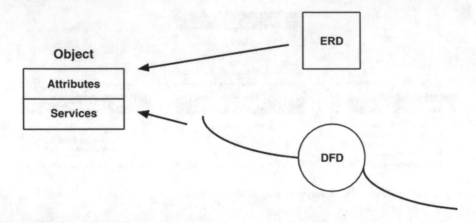

Fig. 3.19 The relationships between an object and the ERD and DFD

Thus, the functional primitive DFDs and the ERD resulting from the normalization process provide the vehicles for providing an object's attributes and services.

3.8 Object-Oriented Databases

There was a movement in the industry to replace the traditional Relational Database Management Systems (RDBMS) with the Object-Oriented Database Management System (OODBMS). Object databases differ greatly from the relational model in that the object's attributes and services are stored together. Therefore, the concept of columns and rows of normalized data becomes extinct. The proponents of OODBMS saw a major advantage in that object databases could also keep graphical and multimedia information about the object, something that relational databases cannot do. The result has been the ultimate creation of different data storages, many of which do not require row and column architecture, but it is expected that the relational model will continue to be used for some time. However, most RDBMS products will become more OO. This means they will use the relational engine but employ more OO capabilities, that is, build a relational hybrid model. In either case, analysts should continue to focus on the logical aspects of capturing the requirements. Changes in the OO methodologies are expected to continue with the evolution of block chain architectures as well as the use of unformatted data using Natural Language methodologies.

3.9   Designing Distributed Objects Using Use-Case Analysis and Design

Use-Cases were first proposed in 1986 as a result of the popularity of the object-oriented paradigm. Use-Case today is widely used in the development of web-based systems and is the appropriate methodology to use for mobile IoT application development. Use-Case was designed to be very effective when defining current and potential actions of a product. That is, Use-Case can be used to model activities that may never have occurred in a system, but are technically possible. Indeed, many system deficiencies occur because a user tried to perform something for the first time. These types of situations are sometimes referred to as supplementary specifications. In many ways, Use-Case methodology represented the next generation of the State Transition Diagrams (STD) discussed earlier in this chapter.

3.9.1   Use-Case Model

A Use-Case model contains three essential components: use-cases, actors, and relationships (Bittner and Spence 2003).

3.9.2   Actors

An Actor represents a user of the system. When interfacing with the system, an "user" can be internal (traditional), consumer, or another system. They are notated using the symbol in Fig. 3.20.

3.10   Use Case

A Use Case identifies a particular interface or "use" that an Actor does with the system to achieve a need. In many ways, the sum of all use cases represents an inventory of all possible transactions and events that can be accomplished with the system. It, in effect, replicates all possible permutations that can occur. In its most decomposed form, each use case defines one transaction. A use case must result in some form of output. Obviously, use cases may have restrictions; certain possible actor requests may require certain authorizations. A use case is denoted by a sphere symbol as shown in Fig. 3.21 (note its similarity with a DFD process).

Figure 3.22 shows a basic Actor/Use Case diagram.

Note that the Use Case model in Fig. 3.22 actually contains two transactions. It could be decomposed to two separate Use Case models as shown in Fig. 3.23.

The third component of the Use-Case modeling is the relationship specified by a data flow line, which often has an arrow to depict directionality. Similar to a DFD, a data flow carries data that will be transformed by the use case process sphere.

Fig. 3.20 Use-case actor symbol

Directionality depicts whether the data is being supplied by the Actor or received by the Actor from the use case process, or both! These relationship data flows are shown in Figs. 3.22 and 3.23.

While a use-case model has three essential symbols, there is another component that is critical. Some analysts call this the description, however, the concept again originated from the DFD in which the actual algorithm inside the process is called a process specification. Process specifications typically contained two forms of description: (1) the actual algorithm in a form of pseudocode, or (2) Pre–Post Conditions. Both can be used together depending on the complexity of the process. Many analysts define a process specification as everything else about the process not already included in the other modeling tools. Indeed, it must contain the

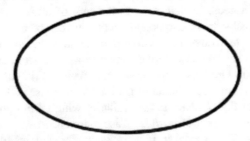

Fig. 3.21 Use case symbol

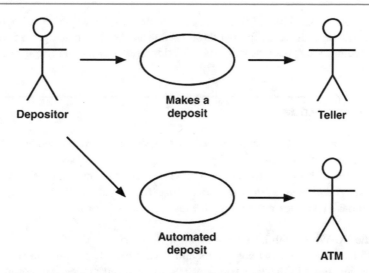

Fig. 3.22 Actor/use case flow

remaining information that normally consists of business rules and application logic. DeMarco suggested that every functional primitive DFD point to a "Minispec" which would contain that process' application logic.[7] We will follow this rule and expand on the importance of writing good application logic even in a Use Case.

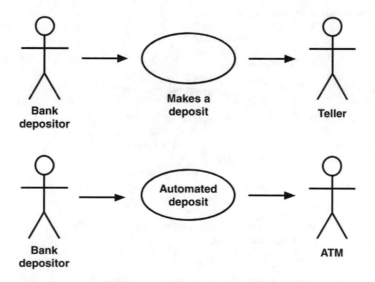

Fig. 3.23 Use case as functional primitives

---

[7]DeMarco (2002).

There are, of course, different styles, and few textbooks that explain the importance to the analyst of understanding how these need to be developed and presented. Like other modeling tools, each process specification style has its good, bad and ugly.

3.11   Pseudocode

The most detailed and regimented process specification is pseudocode or "Structured English". Its format is designed to require the analysts to have a solid understanding of how to write algorithms. The format is very "COBOL-like" and was initially designed as a way of writing functional COBOL programming specifications. The rules governing pseudocode are as follows:

- Use the Do While with an Enddo to show iteration
- Use If-Then-Else to show conditions and ensure each If has an End-If
- Be specific about initializing variables and other detail processing requirements.

Pseudocode is designed to give the analyst tremendous control over the design of the code. Take the following example:

There is a requirement to calculate a 5% bonus for employees who work on the 1st shift and a 10% bonus for workers on the 2nd or 3rd shift. Management is interested in a report listing the number of employees who receive a 10% bonus. The process also produces the bonus checks.

The pseudocode would be:

```
Initialize 10% counter = 0
Open Employee Master File
DoWhile more records
        If Shift = "1" then
                Bonus = Gross_Pay * .05
        Else
                If Shift = "2" or "3" then
                        Bonus = Gross_Pay * .10
                        Add 1 to Counter
                Else
                        Error Condition
        Endif
                Endif
Enddo
Print Report of 10% Bonus Employees
Print Bonus Checks
End
```

The above algorithm gives the analyst great control over how the program should be designed. For example, note that the pseudocode requires that the programmer have an error condition should a situation occur where a record does not contain a 1st, 2nd or 3rd shift employee. This might occur should there be a new shift that was not communicated to the information systems department. Many programmers might have omitted the last "If" check as follows:

```
Initialize 10% counter = 0
Open Employee Master File
DoWhile more records
     If Shift = "1" then
               Bonus = Gross_Pay * .05
     Else
               Bonus = Gross_Pay * .10
               Add 1 to Counter
     Endif
Enddo
Print Report of 10% Bonus Employees
Print Bonus Checks
End
```

The above algorithm simply assumes that if the employee is not on the 1st shift then they must be either a 2nd or 3rd shift employee. Without this being specified by the analyst, the programmer may have omitted this critical logic which could have resulted in a 4th shift worker receiving a 10% bonus! As mentioned earlier, each style of process specification has its advantages and disadvantages, in other words, the good, the bad, and the ugly.

The Good:

The analyst who uses this approach has practically written the program, and thus the programmer will have very little to do with regards to figuring out the logic design.

The Bad:

The algorithm is very detailed and could take a long time for the analyst to develop. Many professionals raise an interesting point: Do we need analysts to be writing process specifications to this level of detail? In addition, many programmers may be insulted and feel that an analyst does not possess the skill-set to design such logic.

The Ugly:

The analyst spends the time, the programmers are not supportive and the logic is incorrect. The result here will be the "*I told you so*" remarks from programmers, and hostilities may grow over time.

3.11.1 Case

Case[8] is another method of communicating application logic. Although the technique does not require as much technical format as pseudocode, it still requires the analyst to provide a detailed structure to the algorithm. Using the same example as in the pseudocode discussion, we can see the differences in format:

```
Case 1st Shift
        Bonus = Gross_Pay * .05
Case 2nd or 3rd Shift
        Bonus = Gross_Pay * .10
        Add 1 to 10% Bonus Employees
Case Neither 1st, 2nd or 3rd Shift
        Error Routine
EndCase
Print Report of 10% Bonus Employees
Print Bonus Checks
End
```

The above format provides control as it still allows the analyst to specify the need for error checking; however, the exact format and order of the logic is more in the hands of the programmer. Let's now see the good, bad and ugly of this approach:

The Good:

The analyst has provided a detailed description of the algorithm without having to know the format of logic in programming. Because of this advantage, CASE takes less time than pseudocode.

The Bad:

Although this may be difficult to imagine, the analyst may miss some of the possible conditions in the algorithm, such as forgetting a shift! This happens because the analyst is just listing conditions as opposed to writing a specification. Without formulating the logic as we did in pseudocode, the likelihood of forgetting or overlooking a condition check is increased.

The Ugly:

Case logic can be designed without concern for the sequence of the logic, that is, the actual progression of the logic as opposed to just the possibilities. Thus the logic can become more confusing because it lacks actual progressive structure. As stated previously, the possibility of missing a condition is greater because the analyst is not actually following the progression of the testing of each condition. There is thus a higher risk of the specification being incomplete.

---

[8]The Case method should not be confused with CASE (Computer Aided Software Engineering) products, which is software used to automate and implement modeling tools and data repositories.

3.12 Pre–Post Conditions

Pre-Post is based on the belief that analysts should not be responsible for the details of the logic, but rather for the overall highlights of what is needed. Therefore, the pre-post method lacks detail and expects that the programmers will provide the necessary details when developing the application software. The method has two components: Pre-Conditions and Post-Conditions. Pre-conditions represent things that are assumed true or that must exist for the algorithm to work. For example, a pre-condition might specify that the user must input the value of the variable X. On the other hand, the post-condition must define the required outputs as well as the relationships between calculated output values and their mathematical components. Suppose the algorithm calculated an output value called Total_Amount. The post-condition would state that Total_Amount is produced by multiplying Quantity times Price. Below is the pre-post equivalent of the Bonus algorithm:

> *Pre-Condition 1:*
>> Access Employee Master file and where 1st shift = "1"
>
> *Post-Condition 1:*
>> Bonus is set to Gross_Pay * .05.
>> Produce Bonus check.
>
> *Pre-Condition 2:*
>> Access Employee Master file and where 2nd shift = "2" or 3rd shift ="3"
>
> *Post-Condition 2:*
>> Bonus is set to Gross_Pay * .10
>> Add 1 to 10% Bonus count.
>> Produce Bonus check and Report of all employees who receive 10% bonuses.
>
> *Pre-Condition 3:*
>> Employee records does not contain a shift code equal to "1", "2", or "3"
>
> *Post-Condition 3:*
>> Error Message for employees without shifts = "1", "2", or "3"

As we can see, the above specification does not show how the actual algorithm should be designed or written. It requires the programmer or development team to find these details and implement the appropriate logic to handle it. Therefore, the analyst has no real input into the way the application will be designed or the way it functions.

The Good:

The analyst need not have technical knowledge to write an algorithm and does not need to spend an inordinate amount of time to develop what is deemed a programming responsibility. Therefore, less technically oriented analysts can be involved in specification development.

The Bad:

There is no control over the design of the logic, and thus the opportunity for misunderstandings and errors is much greater. The analyst and the project are much more dependent on the talent of the development staff.

The Ugly:

Perhaps we misunderstand the specification. Since the format of pre-post conditions is less specific, there is more room for ambiguity.

3.13   Matrix

A matrix or table approach is one that shows the application logic in tabular form. Each row reflects a result of a condition, with each column representing the components of the condition to be tested. The best way to explain a matrix specification is to show an example as shown in Fig. 3.24.

Although this is a simple example that uses the same algorithm as the other specification styles, it does show how a matrix can describe the requirements of an application without the use of sentences and pseudocode.

The Good:

The analyst can use a matrix to show complex conditions in a tabular format. The tabular format is preferred by many programmers because it is easy to read, organized and often easy to maintain. Very often the matrix resembles the array and table formats used by many programming languages.

The Bad:

It is difficult, if not impossible, to show a complete specification in matrices. The above example supports this, in that the remaining logic of the bonus application is not shown. Therefore, the analyst must integrate one of the other specification styles to complete the specification.

Bonus Percent	Shift to be tested
5 % Bonus	1st Shift
10% Bonus	2nd Shift
10% Bonus	3rd Shift

Fig. 3.24 Sample matrix specification

The Ugly:

Matrices are used to describe complex condition levels, where there are many "If" conditions to be tested. These complex conditions often require much more detailed analysis than shown in a matrix. The problem occurs when the analyst, feeling the matrix may suffice, does not provide enough detail. The result: conditions may be misunderstood by the programmer during development.

Conclusion:

The question must be asked again: What is a good specification? We will continue to explore this question. In this chapter we have examined the logic alternatives. Which logic method is best? It depends! We have seen from the examples that each method has its advantages and shortcomings. The best approach is to be able to use them all, and to select the most appropriate one for the task at hand. To do this effectively means clearly recognizing where each style provides a benefit for the part of the system you are working with, and who will be doing the development work. The table below attempts to put the advantages and shortcomings into perspective.

3.14 Problems and Exercises

1. What is an Object?
2. Describe the relationship between a Method and a Service.
3. What is a Class?
4. How does the Object Paradigm change the approach of the analyst?
5. Describe the two types of objects and provide examples of each type.
6. What are Essential Functions?
7. What is an Object Type and how is it used to develop specific type of classes?
8. What is meant by Object and Class Inheritance?
9. What are the association differences between an ERD and an Object diagram?
10. How does functional decomposition operate with respect to classes and objects?
11. What is Coupling and Cohesion? What is their relationship with each other?
12. How does the concept of cohesion relate the structured approach to the object model?
13. What four methods can be used to design a cohesive object?
14. What are Object Databases?
15. What is Client/Server?
16. How do objects relate to Client/Server design?
17. Why is there a need for a hybrid object in Client/Server design?
18. What is Use Case analysis and design?
19. What is meant by distributed objects?

3.15 Mini-project

You have been asked to automate the Accounts Payable process. During your interviews with users you identify four major events as follows:

I. *Purchase Order Flow*

1. The Marketing Department sends a Purchase Order (P.O.) form for books to the Accounts Payable System (APS).
2. APS assigns a P.O. # and sends the P.O.-White copy to the Vendor and files the P.O.-Pink copy in a file cabinet in P.O.#.sequence.

II. *Invoice Receipt*

1. A vendor sends an invoice for payment for books purchased by APS.
2. APS sends invoice to Marketing Department for authorization.
3. Marketing either returns invoice to APS approved or back to the vendor if not authorized.
4. If the invoice is returned to APS it is matched up against the original P. O.-Pink. The PO and vendor invoice are then combined into a packet and prepared for the voucher process.

III. *Voucher Initiation*

1. APS receives the packet for vouchering. It begins this process by assigning a voucher number.
2. The Chief Accountant must approve vouchers > $5,000.
3. APS prepares another packet from the approved vouchers. This packet includes the P.O.-Pink, authorized invoice and approved voucher.

IV. *Check Preparation*

1. Typist receives the approved voucher packet and retrieves a numbered blank check to pay the vendor.
2. Typist types a two-part check (blue, green) using data from the approved voucher and enters invoice number on the check stub.
3. APS files the approved packet with the Check–green in the permanent paid file.
4. The check is either picked up or mailed directly to the vendor.

Assignment:

1. Provide the DFDs for the four events. Each event should be shown as a single DFD on a separate piece of paper.
2. Level each event to its functional primitives.
3. Develop the Process Specifications for each functional primitive DFD.

References

Bittner, K., & Spence, I. (2003). *Use case modeling*. Boston, MA: Addison-Wesley.
Booch, G. (1995). *Object solutions: Managing the object-oriented project* (p. 305).
DeMarco, Tom. (2002). *Structured analysis and system specification* (pp. 85–86).
Langer, A. M. (1997). *The art of analysis*. New York: Springer.
Martin, J., & Odell, J. (1994). *Object oriented methods* (p. 158).

Distributed Client/Server and Data

4

4.1 Client/Server and Object-Oriented Analysis

Client/Server provides another level of sophistication in the implementation of systems. The concept of Client/Server is based on distributed processing, where programs and data are placed in the most efficient places. Client/server systems are typically installed on Local Area Networks (LANs) or Wide Area Networks (WANs). LANs can be defined as multiple computers linked together to share processing and data. WANs are linked LANs. For purposes of this book, we will restrict our discussion about Client/Server within the concepts of application development moving to cloud and mobile environments.

Before you can design effective Client/Server applications for mobility, the organization should commit to the object paradigm. Based on an OO implementation, Client/Server essentially requires one more step: the determination of what portions of an object or class should be moved to client only activities, server only activities, or both across vast mobile networks. Many existing client/server applications need to be expanded to operate in a much more distributed design and one that is not hierarchical.

4.2 Definition of Client/Server Applications

We have already stated that Client/Server is a form of distributed processing. Client/Server applications have three components: a client, a server and a network. Setting aside the implications of the network for a moment, let us understand what clients and servers do. Although Client/Server applications tend to be seen as either permanent client or permanent server programs, we will see that this is not true in the object paradigm.

© Springer Nature Switzerland AG 2020
A. M. Langer, *Analysis and Design of Next-Generation Software Architectures*,
https://doi.org/10.1007/978-3-030-36899-9_4

A "server" is something that provides information to a requester. There are many Client/Server configurations that have permanent hardware servers. These hardware servers typically contain databases and application programs that provide services to requesting network computers (as well as other LANs). This configuration is called "back-end" processing. On the other hand, we have network computers that request the information from servers. We call these computers "clients" and categorize this type of processing as "front-end." When we expand these definitions to applications only, we look at the behavior of an object or class and categorize it as client (requesting services), server (providing services), or both (providing and requesting services).

Understanding how objects become either permanent servers or clients is fairly straightforward. For example, the Cars object in the Car Transmission Types class is categorized as a server. If this were the only use of cars, then it would be called a "dedicated" server object. On the same basis, the Cars object in the Transportation Vehicles class is categorized as a client object. In turn, if it were the only use of the object in a class, it would be defined as a "permanent" client. However; because it exists in more than one class and is polymorphic, the Cars object is really both a client and a server, depending on the placement and behavior of the object. Therefore, when we talk about an object's Client/Server behavior we must first understand the "instance" it is in and the class it is operating within.

The difficulty in Client/Server is in the further separation of attributes and services for purposes of performance across a network. This means that the server services and attributes components of the Cars object might need to be separated from the client ones and permanently placed on a physical server machine.

The client services and attributes will be then be stored on a different physical client machine(s). To put this point into perspective, an object may be further functionally decomposed based on processing categorization. Therefore, the analyst must be involved in the design of the network and must understand how the processing will be distributed across the network. Client/Server analysis should employ Rapid Application Development (RAD)[1] because both analysis and design are needed during the requirements phase of the system. Once the analyst understands the layout of the network, then further decomposition must be done to produce hybrid objects. These hybrid objects break out into dedicated server and object functions as shown below (Fig. 4.1).

Moving to Client/Server is much easier if OO has been completed. Getting the analysis team involved in network design early in the process is much more difficult. The role of the analyst in Client/Server will continue to expand as the distribution of objects in these environments continues to grow and mature.

---

[1]RAD is defined as "an approach to building computer systems which combines the Computer-Assisted Software Engineering (CASE) tools and techniques, user-driven prototyping, and stringent project delivery time limits into a potent, tested, reliable formula for top-notch quality and improvement." Kerr (1994).

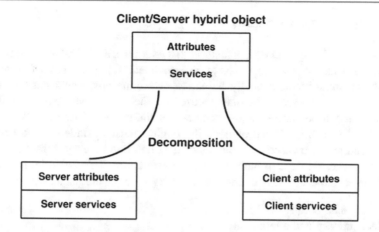

Fig. 4.1 Decomposition of Client/Server objects to dedicated client and server objects

4.3   Introduction to Databases

Chapter 3 focused on application specifications as they relate to process. Using DFDs, PDFs, ERDs, etc., I showed how data elements are defined in the DD. However, the process of completing the DD and building complex relational databases has further steps. This chapter focuses on how to design databases for use with ecommerce Web applications. The completion of the DD and the creation of the database schematic, called the Entity Relational Diagram, provide developers with the data architecture component of the system. We call the process of creating this architecture Logic Data Modeling. The process of logic data modeling not only defines the architecture, it also provides the construct for the actual database, often called the physical database. The physical database differs from its logical counterpart in that it is subject to the rules and formats of the database product that will be used to implement the system. This means that if Oracle is used to implement the logical schema, the database must conform to the specific proprietary formats that Oracle requires. Thus, the logical model provides the first step in planning for the physical implementation. First, I will examine the process of building the appropriate schematic. Even if a packaged software product is selected, the chances are that it will need to use a database product like Oracle. Thus, many of the analysis and design below will be extremely important in determining the best fit for a package.

4.4   Logic Data Modeling

Logic Data Modeling (LDM) is a method that examines a particular data entity and determines what data elements need to be associated with it. There are a number of procedures, some mathematically based, to determine how and what the analyst needs to do. Therefore, LDM only focuses on the stored data with the intent to design what can be defined as the "engine" of the system. Often this "engine" is called the "backend." The design of the engine must be independent from the process and must be based on the rules of data definition theory. Listed below are the eight suggested steps to build the database blueprint. This blueprint is typically called the schema, which is defined as a logical view of the database.

1. identify data entities
2. select primary and alternate keys
3. determine key business rules
4. apply normalization to 3rd normal form
5. combine user views
6. integrate with existing data models (e.g., legacy interfaces)
7. determine domains and triggering operations
8. de-normalize as appropriate.

Prior to providing concrete examples, it is necessary to define the database terms used in this chapter. Below are the key concepts and definitions:

- Entity: an object of interest about which data can be collected. Larson and Larson (2000) define an entity as "a representation of a real-world person, event, or concept." For example, in an ecommerce application, customers, products, and suppliers might be entities. The chapter will provide a method of determining entities from the DFD. An entity can have many data elements associated with it, called attributes.
- Attribute: data elements are typically called attributes when they are associated with an entity. These attributes, or cells of an entity, belong to or "depend on" the entity.
- Key: a key is an attribute of an entity that uniquely identifies a row. A row is defined as a specific record in the database. Therefore, a key is an attribute that has a unique value that no other row or record can have. Typical key attributes are "Social Security Number," "Order Number," etc.
- Business Rule: this is a rule that is assumed to be true as defined by the business. Business rules govern the way keys and other processes behave within the database.
- Normalization: a process that eliminates data redundancy and ensures data integrity in a database.

- User View: the definition of the data from the perspective of the user. This means that how a data element is used, what is its business rules, and whether it is a key or not, depends largely on the user's definition. It is important that analysts understand that data definitions are not universal.
- Domains: this relates to a set of values or limits of occurrences within a data element or attribute of an entity. An example of a domain would be STATE, where there is a domain of 50 acceptable values (i.e., NY, NJ, CA, etc.).
- Triggers: these are stored procedures or programs that are activated or triggered as a result of an event at the database level. In other words, an event (insert, delete, update) may require that other elements or records be changed. This change would occur by having a program stored by the database product (such as Oracle) automatically execute and update the data.
- Cardinality: this concept defines the relationship between two entities. This relationship is constructed based on the number of occurrences or associations that one entity has with another. For example, one customer record could have many order records. In this example, both customer and orders are separate entities.
- Legacy Systems: these are existing applications that are in operation. Legacy applications sometimes refer to older and less sophisticated applications that need to be interfaced with newer systems or replaced completely (see Chap. 10).
- Entity Relational Diagram: a schematic of all the entities and their relationships using cardinal format. An entity relational diagram provides the blueprint of the data, or the diagram of the data engine.

4.5   Logic Data Modeling Procedures

The first step in LDM is to select the entities that will be used to start the normalization process. If DFDs have been completed in accordance with the procedures outlined in Chap. 3, then all data stores that represent data files become transformed into data entities. This approach offers the major advantage of modeling process before data. If DFDs or some comparable process tool is not used, then analysts must rely on the information they can obtain from the legacy systems, such as existing data files, screens, and reports. The following example depicts how a data store from a DFD becomes an entity. The data contained in the data store called "Orders" is represented as an actual form containing many data elements (Fig. 4.2). Thus, this example represents a physical form translated into an LE called a data store, which then is transformed again into an entity (Fig. 4.3).

John's Parts, Inc
1818 West Way
New York, NY 10027

ORDER

ORDER NO: 12345
DATE: November 19, 2019

To: :

    A. Langer & Assoc., Inc.
    John St
    Third Floor
    White Plains, NY, 10963

P.O. NUMBER	DATE SHIPPED	SHIPPED VIA	REQUIRED DATE	TERMS
4R32	3/28/2001	UPS	4/1/2001	30 days

QUANTITY	ITEM ID	ITEM NAME	UNIT PRICE	AMOUNT
6	31	Wires	6.50	39.00
2	27	Wheel Covers	25.00	50.00
				$ 0.00
				$ 0.00
				$ 0.00
				$ 0.00
				$ 0.00
		SUBTOTAL		$ 89.00
		SALES TAX		
		SHIPPING & HANDLING		$ 5.50
		TOTAL DUE		$ 99.50

Fig. 4.2 Sample customer order form

4.6   Key Attributes

The next step in LDM is to select the primary and alternate keys. A primary key is defined as an attribute that will be used to identify a record or occurrence in an entity. The primary key, like any key attribute contains a unique value. Often there is more than one attribute in an entity that contains unique values. We call an attribute that can be a primary key a candidate key attribute. This simply means that this attribute can serve in the role of the primary key. If there is only one candidate,

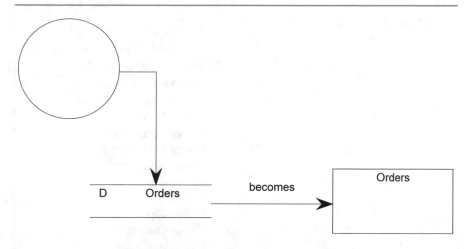

Fig. 4.3 Transition of the order data store into an Entity

then there is no issue: that candidate becomes the primary key. In the event that there is more than one candidate attribute, then one must be selected the primary key, and the others will be called alternate or secondary key attributes. These alternate key attributes provide benefit in the physical database only. This means that they can be used to identify records in the database as an alternative should the primary key not be known. Take the following example. Suppose that an employee entity has two candidate keys: Social-Security-Number and Employee-ID. Employee-ID is selected as the primary key, so Social-Security-Number becomes an alternate key. In the logical entity, Social-Security-Number is treated as any other non-key attribute; however, in the physical database, it can be used (or indexed) to find a record. This could occur when an employee calls to ask someone in Human Resources about accrued vacation time. The Human Resource staff would ask the employee for their Employee-ID. If the employee did not know his/her Employee-ID, the Human Resource staff could ask them for their Social Security Number, and use that information as an alternative way to locate that individual's information. It is important to note that the search on the primary key will be substantially faster, because primary key searches use a method called direct access, as opposed to index methods, which are significantly slower. This raises the question: When there are multiple candidate-key attributes, which key attribute should be selected as the primary key? The answer is the attribute that will be used most often to find the record. This means that Employee-ID was selected the primary key attribute because the users determined that it was the field most often used to locate employee information. Therefore, ecommerce analysts must ensure that they ask users this question during the interview process. Figure 4.4 provides a graphic depiction of the employee entity showing Employee-ID as the primary key attribute and Social-Security-Number as a non-key attribute.

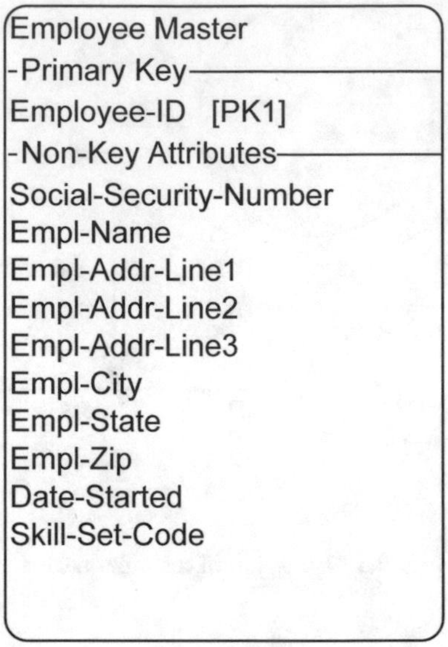

Fig. 4.4 Primary key and
alternate key attributes

```
Employee Master
-Primary Key——————
Employee-ID   [PK1]
-Non-Key Attributes——————
Social-Security-Number
Empl-Name
Empl-Addr-Line1
Empl-Addr-Line2
Empl-Addr-Line3
Empl-City
Empl-State
Empl-Zip
Date-Started
Skill-Set-Code
```

There is another type of key attribute called Foreign keys. Foreign keys provide a way to link tables and create relationships between them. Since foreign keys are created during the process of Normalization, I will defer discussion about them to the section on Normalization in this chapter.

4.7 Normalization

While the next step in LDM is to determine key business rules, it is easier to explain the process of Normalization first. That is, Normalization occurs after Defining Key Business Rules in practice, but not when introducing the topic for educational purposes. Therefore, Key Business Rules will be discussed after Normalization.

Normalization, without question, is the most important aspect of LDM. As mentioned above, Normalization is defined as the elimination of redundancies in an entity and ensures data integrity. It is the latter point that is critical in understanding the value of Normalization in the design of ecommerce database systems. Understanding of the LDM process depends largely on understanding how to implement the Normalization process.

Normalization is constructed in a number of "Normal Forms." While there are five published Normal Forms, Normal Forms 4 and 5 are difficult to implement and most professionals avoid them. Therefore, this book omits Normal Forms 4 and 5. The three Normal Forms of Normalization are listed below. Note that a Normal Form is notated as "NF."

1st NF: No repeating non-key attributes or group of non-key attributes.
2nd NF: No partial dependencies on a part of a concatenated key attribute.
3rd NF: No dependencies of a non-key attribute on another non-key attribute.

Each Normal Form is dependent on the one before it, that is, the process of completing Normalization is predicated on the sequential satisfaction of the Normal Form preceding it. Normalization can be best explained by providing a detailed example. Using the Order form provided in Fig. 4.1, we can start the process of Normalization. Figure 4.5 shows the Logical Equivalent of the Order form in entity format. In this example, the primary key is Order-Number (signified by the "PK" notation), which requires that every order have a unique Order-Number associated with it. It should also be noted that a repeating group made up of five attributes is shown in a separate box. This repeating group of attributes correlates to an area on the Order form, which often is referred to as an order line item. This means that each item associated with the order appears in its own group, namely the item identification, its name, unit price, quantity, and amount. The customer order in Fig. 4.1 shows two items associated with the Order-Number 12345.

The process of determining compliance with Normalization is to evaluate whether each normal form or NF has been satisfied. This can be accomplished by testing each NF in turn. Thus, the first question to ask is: Are we in 1st NF? The answer is no because of the existence of the repeating attributes: Item-ID, Item-Name, Quantity, Unit-Price, and Amount, or as specified above an "order line item." In showing this box, the example exposes the repeating group of items that can be associated with a customer order. Another way of looking at this phenomenon is to see that within the Order, there really is another entity, which has its own key identification. Since there is a repeating group of attributes, there is a 1st NF failure. Anytime an NF fails or is violated, it results in the creation of another entity. Whenever there is a 1st NF failure, the new entity will always have as its primary key a concatenated "group" of attributes. This concatenation, or joining of multiple attributes to form a specific value, is composed of the primary key from the original entity (Orders) attached with a new key attribute from the repeating group of elements. The new key must be an attribute that controls the other group of attributes. In this example, the controlling attribute is Item-ID. After the new "key attribute" is determined, it is concatenated with the original key attribute from the Orders entity. The remaining non-key attributes will be removed from the original entity to become non-key attributes of the new entity. This new entity is shown in Fig. 4.6.

The new entity, called Order Items, has a primary key that reflects the concatenation of the original key Order-Number from the entity Orders, combined with Item-ID, which represents the controlling attribute for the repeating group. All of the other repeating attributes have now been transferred to the new entity. The new entity Order Items allows the system to store multiple order line items as required. The original entity left without this modification would have limited the number of occurrences of items artificially. For example, if the analyst/designer had defaulted to five groups of order line items, the database would always have five occurrences

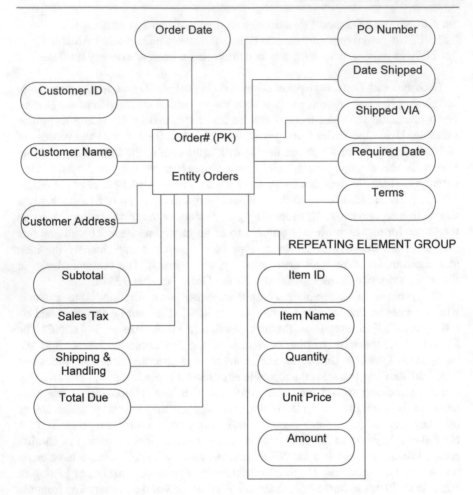

Fig. 4.5 Orders entity and its associated attributes

of the five attributes. If most orders, in reality, had fewer than five items, then significant space would be wasted. More significant is the case where the order has more than five items. In this case, a user would need to split the order into two physical orders so that the extra items could be captured. These two issues are the salient benefits of attaining entities in their 1st NF. Therefore, leaving the entity Order as is would in effect create an integrity problem.

Once the changes to the entity Orders has been completed, and the new entity Order Item has been completed, the system is said to be a database in 1st NF. It is important to note that the new primary key of the entity Order Items is the combination of two attributes. While the two attributes maintain their independence as separate fields of data, they are utilized as one combined value for purposes of their role as a key attribute. For example, based on the data in the Order form from

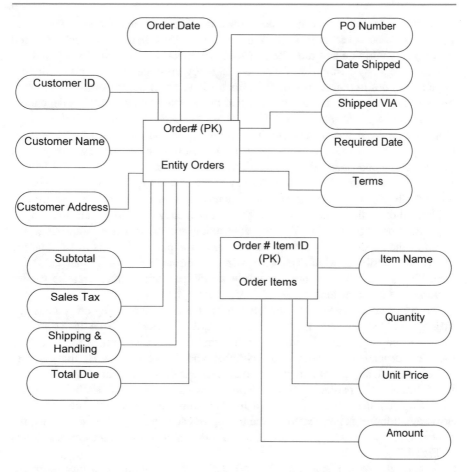

Fig. 4.6 Orders in 1st NF

Fig. 4.1, the entity Order Items would have two records. The first record would have the primary key of 1234531, which would be the concatenation of Order-Number (12345) with Item-ID (31). The second record would be 1234527, which is the same Order-Number, but concatenated with the second Item-ID (27). From an SQL feature perspective, while the key attribute concatenates each attribute into one address, it can be searched as separate fields. So, a user could search for all the items associated with Order 12345, by simply searching on Order Items that contain an Order-Number = "12345." This exemplifies the power of versatility in the relational model. Once 1st NF has been reached, the next test must ensue, that is, testing for compliance with 2nd NF.

Second NF testing applies only to entities that have concatenated keys. Therefore, any entity that is in 1st NF and does not have a concatenated primary key must already be in 2nd NF. In our example, then, the entity "Orders" is already in 2nd NF

because it is in 1st NF and does not have a concatenated primary key. The entity
Order Items, however, is in a different category. Order Items has a concatenated
primary key attribute and must be tested for compliance with 2nd NF. Second NF
requires the analyst to ensure that every non-key attribute in the entity is totally
dependent on all components of the primary key, or all of its concatenated attributes. When we apply the test, we find that the attribute "Item-Name" is dependent
only on the key attribute "Item-ID." That is, the Order-Number has no effect or
control over the name of the item. This condition is considered a 2nd NF failure.
Once again, a new entity must be created. The primary key of the new entity is the
portion of the concatenated key that controlled the attribute that caused the failure.
In other words, Item-ID is the primary key of the new entity, because "Item-Name"
was wholly dependent on the attribute "Item-ID." It is worthwhile at this time to
explain further the concept of attribute dependency. For one attribute to be
dependent on another infers that the controlling attribute's value can change the
value of the dependent attribute. Another way of explaining this is to say that the
controlling attribute, which must be a key, controls the record. That is, if the
Item-ID changes, then we are looking at a different Item Name, because we are
looking at a different Item record.

To complete the creation of the new entity, Items, each non-key attribute in the
original entity Order Items must be tested for 2nd NF violation. Note that as a result
of this testing, "Quantity" and "Amount" stay in the Order Items entity because
they are dependent on both Order-Number and Item-ID. That is, the quantity
associated with any given Order Items occurrence is dependent not only on the Item
itself, but also the particular order it is associated with. We call this being wholly
dependent on the concatenated primary key attribute. Thus, the movement of
non-key attributes is predicated on the testing of each non-key attribute against the
concatenated primary key. The result of this test establishes the three entities shown
in Fig. 4.7.

The results of implementing 2nd NF reflect that without it, a new Item (or
Item-ID) could not have been added to the database without an order. This obviously would have caused major problems. Indeed, the addition of a new Item would
have to precede the creation of that Item with a new Order. Therefore, the new
entity represents the creation of a separate Item master file as shown in Fig. 4.7.

Figure 4.7 represents Orders in 2nd NF. Once again, we must apply the next
test-3rd NF to complete Normalization. Third NF tests the relationship between two
non-key attributes to ensure that there are no dependencies between them. Indeed, if
this dependency were to exist, it would mean that one of the non-key attributes
would, in effect, be a key attribute. Should this occur, the controlling non-key
attribute would become the primary key of the new entity. Testing this against the
sample entity reflects that Customer-Name and Customer-Address[2] are dependent
on Customer-ID. Therefore, the entity Orders fails 3rd NF and a new entity must be
created. The primary key of the new entity is the non-key attribute that controlled

---

[2]Customer-Address would normally be composed of three address lines and the existence of State,
City, and Zip-code. It has been omitted from this example for simplicity.

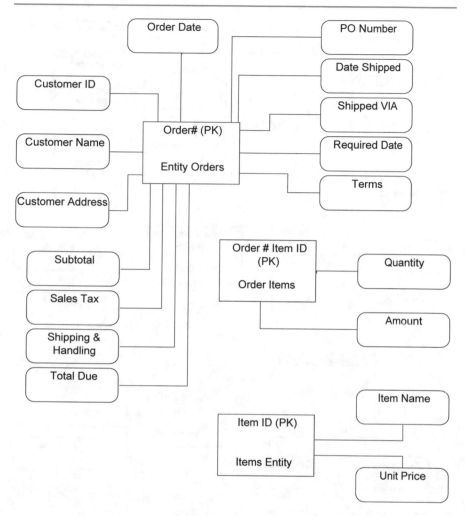

Fig. 4.7 Orders in 2nd NF

the other non-key attributes; in this case Customer-ID. The new entity is called Customers, and all of the non-key attributes that depend on Customer-ID are moved to that entity as shown in Fig. 4.8.

What is unique about 3rd NF failures is that the new key attribute remains as a non-key attribute in the original entity (in this case: Orders). The copy of the non-key attribute Customer-ID is called a foreign key and is created to allow the Order entity and the new Customer entity to have a relationship. A relationship between two entities can exist only if there is at least one common keyed attribute between them. Understanding this concept is crucial to what Normalization is intended to accomplish. Looking at Fig. 4.8, one can see that the entity Order and

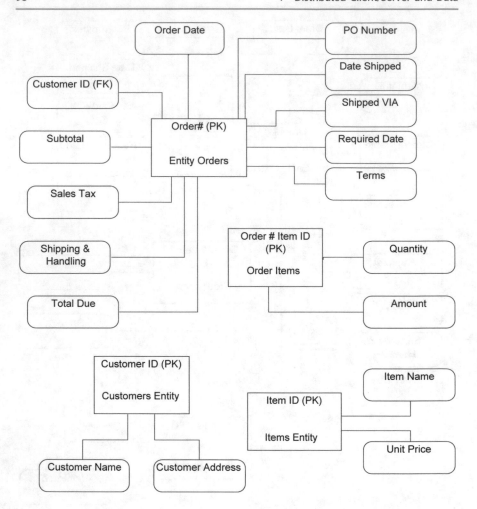

Fig. 4.8 Orders in preliminary 3rd NF

Order Items have a relationship because both entities have a common keyed attribute: Order-ID. The same is true in the creation of the Item entity, which resulted from a 2nd NF failure. The relationship here is between the Order Item entity and the Item entity, where both entities contain the common key attribute Item-ID. Both of these relationships resulted from the propagation of a key attribute from the original entity to the newly formed entity during the normalization process. By propagation, we mean that a pointer, or copy of the key attribute is placed in the new entity. Propagation is implemented using foreign keys and is a natural result of the process. Note that the "PK" is followed by an "FK" signifying that the keyed attribute is the result of a propagation of the original key attribute. Such is not the case in 3rd NF. If Customer-ID were to be removed from the Orders entity, then the relationship between Orders and Customers would not exist because there would be no common keyed attribute between the two entities. Therefore, in 3rd NF, it is necessary to force the relationship because a natural propagation has not occurred. This is accomplished by creating a pointer from a non-keyed attribute to the primary keyed copy, in this case Customer-ID. The concept of a pointer is important. Foreign key structures are typically implemented internally in physical databases using indexes. Indexes, or indirect addresses, are a way of maintaining database integrity by ensuring that only one copy of an attribute value is stored. If two copies of Customer-ID were stored, changing one of them could create an integrity problem between Orders and Customers. The solution is to have the Customer-ID in Orders "point" indirectly to the Customer-ID key attribute in the Customer entity. This ensures that a Customer-ID cannot be added to the Orders entity that does not exist in the Customer master entity.

The question now is whether the entities are in 3rd NF. Upon further review, we see the answer is no! Although it is not intuitively obvious, there are three non-key attributes that are dependent on other non-key attributes. This occurs first in the Order Items entity. The non-key attribute "Amount" is dependent on the non-key attribute "Quantity." Amount represents the total calculated for each item in the order. It is not only dependent on "Quantity," but also dependent on "Unit-Price." This occurs frequently in attributes that are calculations. Such attributes are called derived elements, and are eliminated from the database. Indeed, if we store Quantity and Unit-Price, "Amount" can be calculated separately as opposed to being stored as a separate attribute. Storing the calculation would also cause integrity problems. For example, what would happen if the quantity or unit price would change? The database would have to recalculate the change and update the Amount attribute. While this can be accomplished, and will be discussed later in this chapter, it can be problematic to maintain in the database and cause performance problems in production ecommerce systems. The Orders entity also contains two derived attributes: Subtotal and Total-Due. Again, both of these attributes are removed. The issue is whether the removal of derived attributes should be seen as a 3rd NF failure. Date (2000) views these failures as outside of 3rd NF, but in my view, they represent indirect dependencies on other non-key attributes and should be included as part of the 3rd NF test. In any case, we all agree that derived elements should be removed in the process of LDM. The 3rd NF LDM is modified to reflect the removal of these three attributes as shown in Fig. 4.9.

Once 3rd NF is reached the analyst should create the Entity Relational Diagram (ERD), which will show the relationships or connections among the entities. The relationship between entities is established through associations. Associations define the cardinality of the relationship using what is known as the Crow's Foot Method as shown in Fig. 4.10.

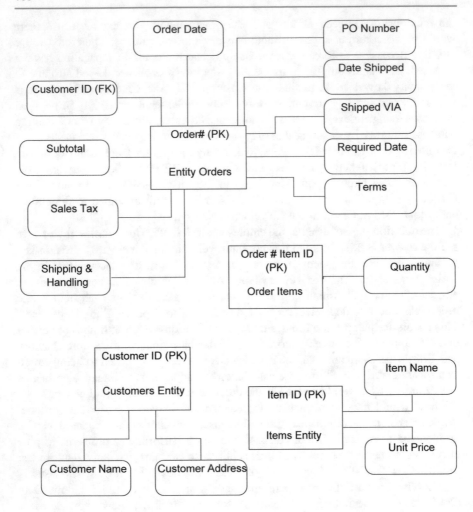

Fig. 4.9 Orders in second phase of 3rd NF

The Crow's Foot Method is only one of many formats. The method contains three key symbols:

≤   denotes the cardinality of many occurrences
o   denotes zero occurrences
| denotes one occurrence.

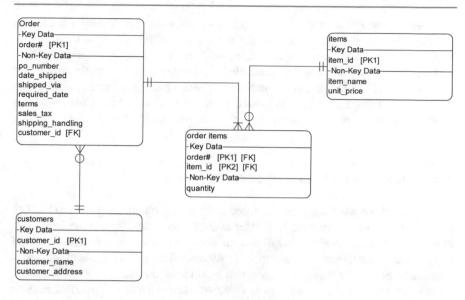

Fig. 4.10 The entities in ERD format using crow's Feet

Therefore, the ERD in Fig. 4.10 depicts the relationships of all the entities as follows:

1. One, and only one (signified by the double lines) Order record can have one to many Order Item records. It also shows that any Order in the Order Items entity must exist in the Order entity.
2. One and only one Item record can have zero to many Order Item records. The difference in this relationship and the one established between Orders and Order Items is that Items may not have a relationship with Order items, signified by the zero in the Crow's Foot. This would often occur when there is a new item that has not yet received any orders.
3. The Order Items entity has a primary key, which is a concatenation of two other primary keys: Order-ID from the Orders entity, and Item-ID from the Items entity. This type of relationship is said to be an "associative" relationship because the entity has been created as a result of a relational problem. This relational problem exists because the Order entity has a "many-to-many" relationship with the Items entity. Thus the 1st NF failure, which created the associative entity Order Items, is really the result of a "many-to-many" situation. A many-to-many relationship violates Normalization because it causes significant problems with SQL coding. Therefore, whenever a many-to-many relationship occurs between two entities, an associative entity is created which will have as its primary key the concatenation of the two primary keys from each entity. Thus, associative entities make many-to-many relationships into two

one-to-many relationships so that SQL can work properly during search routines. Associative entities are usually represented with a diamond box.
4. One and only one Customer can have zero-to-many Orders, also showing that a Customer may exist who has never placed an order. As an example, this would be critical if the business were credit cards, where consumers can obtain a credit card even though they have not made a purchase. Note that the Customer-ID is linked with Orders through the use of a non-key foreign key attribute.

4.8   Limitations of Normalization

Although 3rd NF has been attained, there is a major problem with the model. The problem relates to the attribute Unit-Price in the Items entity. Should the Unit-Price of any Item change, then the calculation of historical Order Item purchases would be incorrect. Remember that the attribute "Amount" was eliminated because it was a derived element. This might suggest that Normalization does not work properly! Such is not the case. First, we need to evaluate whether putting "Amount" back in the ERD would solve the problem. If the Unit-Price were to change, then Amount would need to be recalculated before it was done. While this might seem reasonable, it really does not offer a solution to the problem, just a way around it. The actual problem has little to do with the attribute "Amount", but more to do with a missing attribute. The missing attribute is Order-Item-Unit-Price, which would represent the price at the time of the order. Order-Item-Unit-Price is dependent on both the Order and the Item and therefore would become a non-key attribute in the Order Items entity (i.e., it is wholly dependent on the entire concatenated primary key). The only relationship between Unit-Price and Order-Item-Unit-Price is at the time the order is entered into the system. In this situation, an application program would move the value or amount of the Unit-Price attribute into the Order-Item-Unit-Price attribute. Thereafter, there is no relationship between the two attributes. Because this is a new data element that has been discovered during Normalization, it must be entered into the Data Dictionary. Thus, a limitation of Normalization is that it cannot normalize what it does not have; it can normalize only the attributes that are presented to the formula. However, the limitation of Normalization is also an advantage: the process can help the analyst recognize that a data element is missing. Therefore, Normalization is a "data-based" tool that the analyst can use to reach the Logical Equivalent. Figure 4.11 shows the final ERD with the addition of Order-Item-Unit-Price.

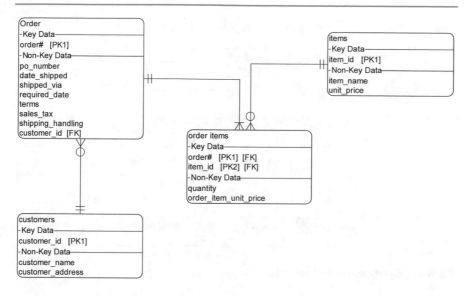

Fig. 4.11   Final ERD with Order-Item-Unit-Price

4.9   The Supertype/Subtype Model

A troublesome database issue occurs in the LDM when there are records within an entity that can take on different characteristics or have many "types" of attributes. "Type" means that a portion of the attributes in a specific record can vary depending on the characteristic or identification of the row within that entity. Another way of defining type is to describe it as a group of attributes within a given record that are different from other records of the same entity depending on the type of record it represents. This type is referred to as a "subtype" of the record. A subtype, therefore, is the portion of the record that deviates from the standard or "supertype" part of the record. The "supertype" portion is always the same among all the records in the entity. In other words, the "supertype" represents the global part of the attributes in an entity. The diagram below in Fig. 4.12 depicts the supertype/subtype relationship.

The difference between a subtype and an ordinary type identifier (using a foreign key) is the occurrence of at least one non-key attribute that exists only in that subtype record. The major reason to create a supertype/subtype relationship is the occurrence of multiple permutations of these unique attributes that exist in just certain subtype records. Limiting these permutations of attributes within one record format can be problematic. First, it can waste storage, especially if each subtype has significant numbers of unique attributes. Second, it can create significant performance problems particularly with the querying of data. Using Fig. 4.12, we can see two ways to store this data. The first (Fig. 4.13) is a basic representation where all

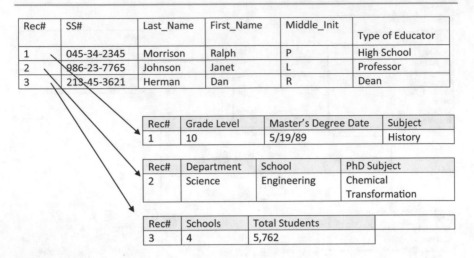

Rec#	SS#	Last_Name	First_Name	Middle_Init	Type of Educator
1	045-34-2345	Morrison	Ralph	P	High School
2	986-23-7765	Johnson	Janet	L	Professor
3	213-45-3621	Herman	Dan	R	Dean

Rec#	Grade Level	Master's Degree Date	Subject
1	10	5/19/89	History

Rec#	Department	School	PhD Subject
2	Science	Engineering	Chemical Transformation

Rec#	Schools	Total Students	
3	4	5,762	

Fig. 4.12 Supertype/subtype relationship

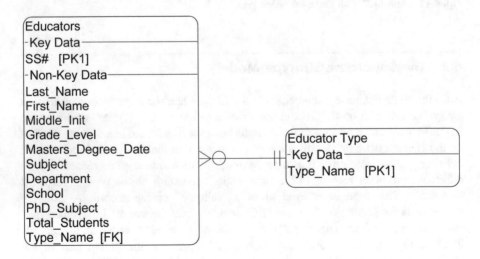

Fig. 4.13 Educator ERD using Foreign Key Identifier

the permutations exist in one entity called "Educators." The "type" of row is identified by using a foreign key pointer to a validation entity called "Educator Type."

Although this representation of the data uses only one entity, it wastes storage space because all of the attributes of the entity are never needed by any one "type" of record. Furthermore, a user must know which attributes need to be entered for a particular type of record. This method of logic data modeling violates the concepts

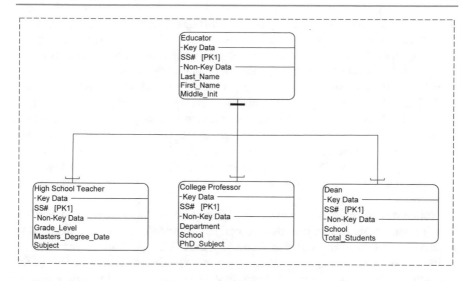

Fig. 4.14 Educator entity supertype/subtype model

of normalization, and entrusts the integrity of values of elements in an entity to either an application program's control (stored procedure), or to the memory of the user. Neither of these choices is particularly dependable or has proven to be a reliable method of data integrity.

On the other hand, Fig. 4.14 provides a different solution using the Supertype/Subtype model.

This model constructs a separate entity for each type of educator, linked via a special notation in the relational model, known as the supertype/subtype relation. The relationship is mutually exclusive, meaning that the supertype entity Educator can have only one of the three subtypes for any given supertype occurrence. Therefore, the relationship of one record in a supertype must be one-to-one with only one subtype. The supertype/subtype model creates a separate subtype entity to carry only the specific attributes unique to its subtype.

There are two major benefits to this entity structure. First, the construct saves storage because it stores only the attributes it needs in each entity. Second, the subtype information can be directly addressed without accessing its related super-type. This is possible because each subtype entity contains the same primary key as its parent. This capability is significant because a user can automatically obtain the unique information from any subtype without having to search first through the supertype entity. This is particularly beneficial when the number of records in each subtype varies significantly. Suppose, for example, there are 6 million educators in the database. The Educator database would therefore contain 6 million rows. Let's say that 5 million of the educators are high school teachers, and as such, the High School subtype entity has 5 million records. Eight hundred thousand educators are professors, and the remaining 200,000 educators are deans; therefore, the Professor database and Dean database have 800,000 and 200,000 records, respectively. Using the supertype/subtype model applications could access each subtype without searching through every record in the database. Furthermore, because access to one subtype does not affect the other, performance is greatly improved.

It is important to note that the Supertype/Subtype model is not limited to mutual exclusivity, that is, it can support multiple subtype permutations. For example, suppose an educator could be a high school teacher, college professor, and a dean at the same time, or any permutation of the three types. The sample model would then be modified to show separate one-to-one relationships as opposed to the "T" relationship shown in Fig. 4.14. The alternative model is represented in Fig. 4.15.

Supertype/Subtypes can cascade, that is, they can continue to iterate or decompose within each subtype. This is represented in Fig. 4.16.

Notice that in the above example the same primary key continues to link the "one-to-one" relationships between the entities. In addition, Fig. 4.16 also shows another possibility in the supertype/subtype model. This possibility reflects that a subtype can exist without containing any non-key attributes. This occurs in the example in the subtype entity Adjunct Prof. The "empty" entity serves only to identify the existence of the subtype, without having a dedicated non-key attribute associated with it. The Adjunct Prof entity, therefore, is created only to allow the other two subtypes (Tenured Prof and Contract Prof) to store their unique attributes. This example shows how supertype/subtype models can be constructed, and how they often have subtypes that are created for the sole purpose of identification.

Cascading subtypes can mix methods, that is, some levels may not be mutually exclusive, while other cascade levels can be mutually exclusive as shown in Fig. 4.17.

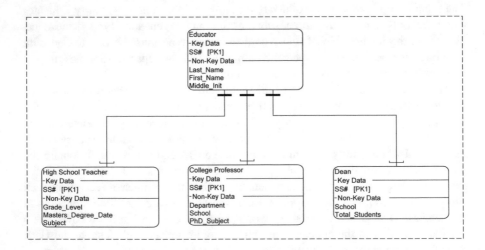

Fig. 4.15 Supertype/subtype model without mutual exclusivity

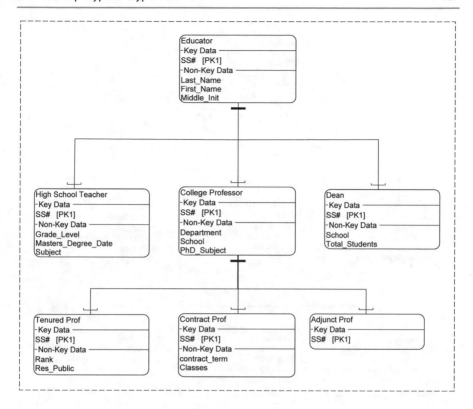

Fig. 4.16 Cascading subtypes

There is a controversial issue among database developers. The controversy relates to whether it is necessary to create a special attribute that identifies which entity contains the subtype entry for any given supertype. In other words, how does the database know which subtype has the continuation information? This dilemma is especially relevant when mutually exclusive relationships exist in the supertype/subtype. The question is ultimately whether the supertype/subtype model needs to contain an identifier attribute that knows which subtype holds the continuation record, or is the issue resolved by the physical database product? Fleming and von Halle addressed this issue in the Handbook of Database Design, where they suggest that the "attribute is at least partially redundant because its meaning already is conveyed by the existence of category or subtype relationships" (p. 162). Still, the issue of redundancy may vary among physical database products. Therefore, I suggest that the logical model contain a subtype identifier for mutually exclusive supertype/subtype relationships as shown in Fig. 4.18.

Note that the above example has the subtype identifier, Professor Types as a validation entity in 3rd normal form.

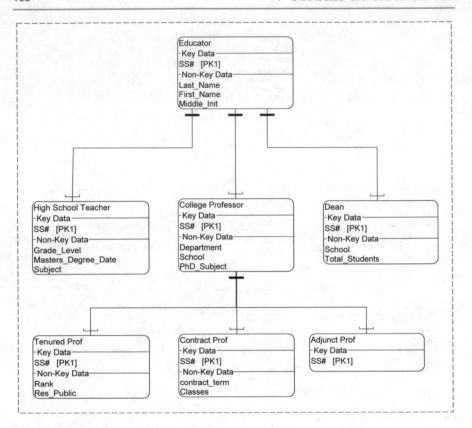

Fig. 4.17 Cascading subtypes with alternating exclusivities

Supertype/subtypes must also be normalized following the rules of Normalization. For example, the subtype Educator Types contains elements that are not in 3rd NF. Attributes Grade_Level and Subject in the subtype entity High School Teacher can be validated using a look-up table. Department, School, and PhD_Subject can also be validated. The resulting 3rd NF ERD is shown below in Fig. 4.19.

4.10   Key Business Rules

Key business rules are the rules that govern the behavior between entities when a row is inserted or deleted. These business rules are programmed at the database level using stored procedures and triggers (see Step 7: determine domains and triggering operations). These procedures are typically notated as constraints. Constraints enforce the key business rules that will be defined by the analysts and are the basis of what is meant by referential integrity, that is, the integrity based on the relations between tables. The process of insertion and deletion focuses on the

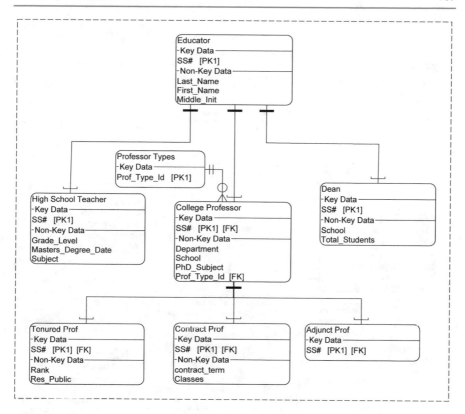

Fig. 4.18 Supertype/subtype with subtype identifier element

relationship between the parent entity and the child entity. A child entity is always the entity that has the Crow's Foot pointing to it. Based on the ERD in Fig. 4.10 the parent-child entity relationships are as follows:

- Orders entity is the parent of Order Items entity (child).
- Customer entity is the parent of Orders entity (child).
- Items entity is the parent of Order Items entity (child).

When discussing insertion of a row, it is always from the perspective of the child entity. That is, key business rules governing the insertion of a child record concern what should be done when attempting to insert a child record that does not have a corresponding parent record. There are six alternatives:

1. Not Allowed: this means that the constraint is to disallow the transaction. For example, in Fig. 4.11, a user could not insert an Order Item (child) for an Order-Number that did not exist in the Orders entity (parent). Essentially, the integrity of the reference would be upheld, until the Order-Number in the Orders entity was inserted first.

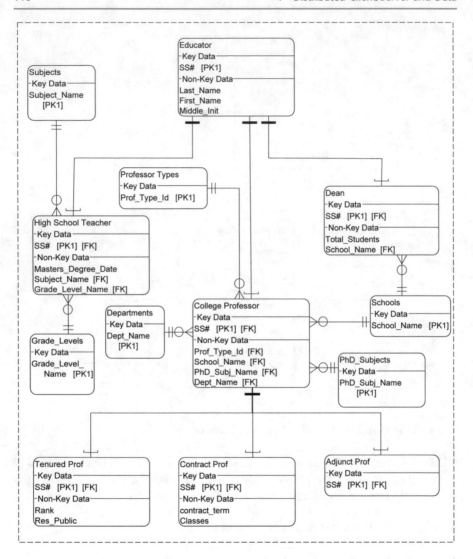

Fig. 4.19 Supertype/subtype in 3rd NF

2. Add Parent: this means that if the parent key does not exist, it will be added at the same time. Using Fig. 4.11, this would mean that the user would be prompted to add the Order-Number to the Orders entity before the child Item would be inserted. The difference between Not Allowed and Add Parent is that the user can enter the parent information during the insertion of the child transaction. Using this rule still enforces referential integrity.
3. Default Value: the use of a default value allows a "dummy" row to be inserted into the parent. An example of the use of a default occurs when collection

agencies receive payments from an unknown person. The parent entity is "Account" and the child entity is "Payments." The Account entity would have a key value called "Unapplied" which would be used whenever an unidentified payment was collected. In this scenario, it is appropriate to have the dummy record because the child transaction is really unknown, but at the same time needs to be recorded in the database. It is also useful because the user can quickly get a list of "unapplied" payments and it upholds referential integrity.

4. Algorithm: an algorithm is an "intelligent" default value. Using the same example as (3), suppose the user wanted to track unapplied payments by State. For example, if an unapplied payment were received in New York, the parent (Account entity) would have a record inserted with a value "Unapplied-New York." Therefore, each State would have its own default. There are also default keys that are based on sophisticated algorithms to ensure that there is an understanding to the selection of the parent's key attribute value. Again, this selection ensures referential integrity because a record is inserted at both the parent and child entities.

5. Null: assigning a null means that the parent does not exist. Most database products such as Oracle allow such selection, and while it is maintained within the product, it violates referential integrity because the parent is unknown.

6. Don't Care: this essentially says that the user is willing to accept that referential integrity does not exist in the database. The user will tell you that they never wish to balance the records in the child with those in the parent. While this happens, it should be avoided, because it creates a system without integrity.

When discussing deletion of a row, it is always from the perspective of the parent entity. That is, key business rules governing the deletion of a parent record concern what should be done when attempting to delete a parent record that has corresponding child records. There are similarly six alternatives:

1. Not Allowed: this means that the constraint is to disallow the deletion of the parent record. In other words, if there are children records, the user cannot delete the parent. For example, if in Fig. 4.11, a user could not delete an Order (parent) if there were corresponding records in the Order Items entity (child). This action would require the user first to delete all of the Order Items or children records before allowing the parent Order to be deleted.

2. Delete All: this is also known as cascading, because the system would automatically delete all child associations with the parent entity. Using the same example as (1), the children records in Order Items would automatically be deleted. While this option ensures referential integrity, it can be dangerous because it might delete records that are otherwise important to keep.

3. Default Value: the use of a default value is the same as in insertion, that is, it allows a "dummy" row to be inserted into the parent. This means that the original parent is deleted, and the child records are redirected to some default value row in the parent entity. This is sometimes useful when there are many old parent records, such as old part-numbers, that are cluttering up the parent

database. If keeping the child records is still important, they can be redirected to a default parent row, such as "Old Part-Number."

4. Algorithm: the use of the algorithm is the same as with an insertion. As in the case of (3) above, the default value might be based on the type of product or year it became obsolete.

5. Null: as in the case of insertion, the assigning of a null means that the parent does not exist. This creates a situation where the child records become "orphans." Referential integrity is lost.

6. Don't Care: same as in insertion. The database allows parent records to be deleted without checking to see if there are corresponding child records in another entity. This also results in losing referential integrity and creates "orphans."

In summary, key business rules are concerned with the behavior of primary keys during insert and delete operations. There are six alternative options within each operation (insert and delete). Four of the options uphold referential integrity, which is defined as the dependability of the relationships between items of data. Data integrity is an issue any time there is change to data, which in ecommerce systems will be frequent. Thus, the ecommerce analyst must ensure that once primary keys have been determined, it is of vital importance that users are interviewed regarding their referential integrity needs. Analysts should not make these decisions in a vacuum and need to present the advantages of referential integrity appropriately to users so that they can make intelligent and well-informed decisions.

This discussion of key business rules was predicated on using examples derived from the discussion on Normalization. As discussed earlier in this section, the application of normalization occurs after the determination of key business rules, especially since it may indeed affect the design of the ERD, and in the programming of stored procedures. This will be discussed further in the Determine Domains and Triggering Operations section of this chapter.

4.11   Combining User Views

The application of Normalization focused on breaking up or decomposing entities to include the correct placement of data. Each NF failure resulted in creating a new entity; however, there are situations where certain entities may need to be combined. This section is labeled "Combining User Views" because the meaning of data is strongly dependent on how the user defines a data element. Unfortunately, there are circumstances where data elements are called different things and defined differently by different users in different departments. The word "different" is critical to the example. In cases where we think we have two entities, we may, in fact have only one. Therefore, the process of combining user views typically results in joining two or more entities as opposed to decomposing them as done with Normalization. The best way to understand this concept is to recall the earlier discussion on Logical Equivalents. This interpretation of the Logical Equivalent will focus on the data rather than the process. Suppose there are two entities created from two different departments. The first department defines the elements for an entity called "Clients" as shown in Fig. 4.20.

The second department defines an entity called "Customers" as shown in Fig. 4.21.

Upon a closer analysis and review of the data element definitions, it becomes apparent that the two departments are looking at the same object. Notwithstanding whether the entity is named Client or Customer, these entities must be combined. The process of combining two or more entities is not as simple as it might sound. In the two examples, there are data elements that are the same with different names, and there are unique data elements in each entity. Each department is unaware of the other's view of the same data, and by applying logical equivalencies the following single entity results as shown in Fig. 4.22.

The above example uses names that made it easier for an analyst to know they were the same data elements. In reality, such may not be the case, especially when working with legacy systems. In legacy systems, names and definitions of elements can vary significantly among departments and applications. Furthermore, the data definitions can vary significantly. Suppose Client is defined as VARCHAR2(35) and Customer as VARCHAR2(20). The solution is to take the larger definition. In still other scenarios, one element could be defined as alphanumeric, and the other numeric. In these circumstances the decisions become more involved with user conversations. In either situation, it is important that the data elements do get

Fig. 4.20 The client entity

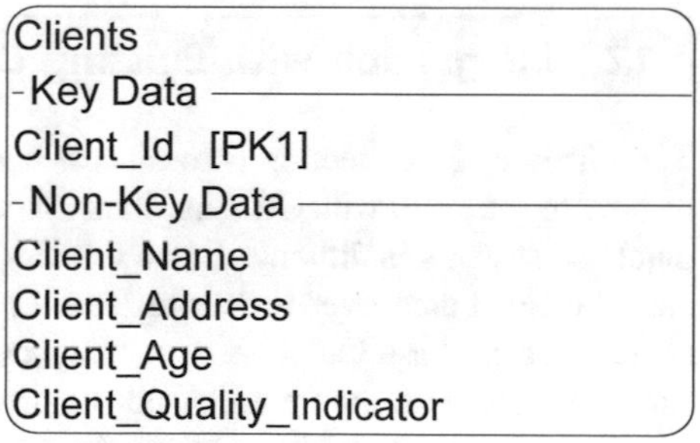

```
Clients
- Key Data ───────────
Client_Id  [PK1]
- Non-Key Data ───────
Client_Name
Client_Address
Client_Age
Client_Quality_Indicator
```

Fig. 4.21 The customer entity

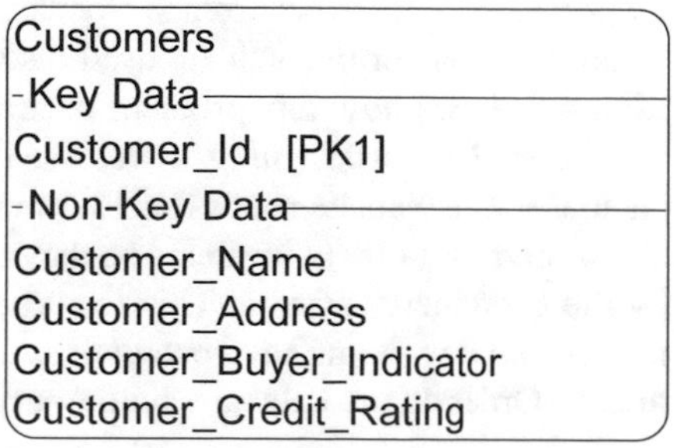

```
Customers
- Key Data────────────
Customer_Id  [PK1]
- Non-Key Data────────
Customer_Name
Customer_Address
Customer_Buyer_Indicator
Customer_Credit_Rating
```

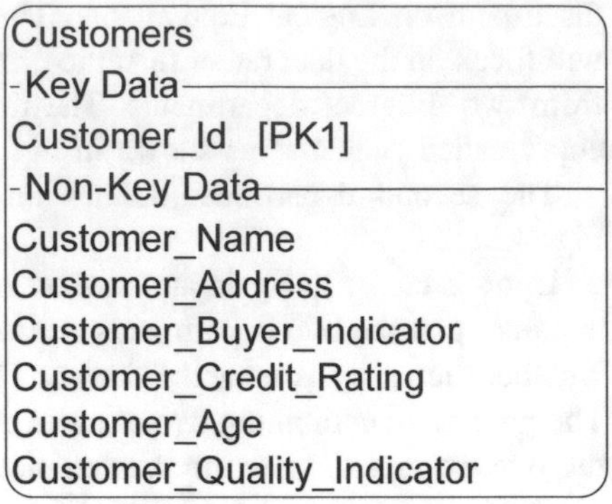

Fig. 4.22 Combined client
and customer entity

```
╭─────────────────────────────────────╮
│ Customers                            │
│ ─Key Data─────────────────────────── │
│ Customer_Id   [PK1]                  │
│ ─Non-Key Data──────────────────────  │
│ Customer_Name                        │
│ Customer_Address                     │
│ Customer_Buyer_Indicator             │
│ Customer_Credit_Rating               │
│ Customer_Age                         │
│ Customer_Quality_Indicator           │
╰─────────────────────────────────────╯
```

combined and that users agree-to-agree. In cases where user agreement is difficult, then analysts can take advantage of a data dictionary feature called Alias. An Alias is defined as an alternate name for a data element. Multiple Aliases can point to the same data dictionary entry. Therefore, screens can display names that are Aliases for another element. This alternative can solve many problems when using different names is necessary.

Another important issue in combining user views is performance. While analysts should not be overly concerned about performance issues during LDM, it should not be ignored either. Simply put, the fewer entities, the faster the performance; therefore, the least number of entities that can be designed in the ERD the better.

4.12   Integration with Existing Data Models

The purpose of this section is to discuss specific analysis and design issues relating to how to integrate with existing database applications. The connectivity with other database systems is difficult. Indeed, many firms approach the situation by phasing each business area over time into a new re-developed operation. In these circumstances, each phased area needs a "Legacy Link" which allows the "old" applications to work with the new phased-in software.

Linking entities with existing databases may force ecommerce analysts to rethink how to preserve integrity while still maintaining the physical link to other corporate data. This occurrence is a certainty with ecommerce systems given that certain portions of the data are used inside and outside the business. The following example shows how this problem occurs:

The analyst is designing a website that utilizes the company's Orders Master database. The website needs this information to allow customers to see information about their past orders for items so they can match it to a product database supplied by the ecommerce system. This feature is provided to customers to allow them to understand how items have been utilized to make their products. Unfortunately, the master Order Items database holds only orders for the past year and then stores

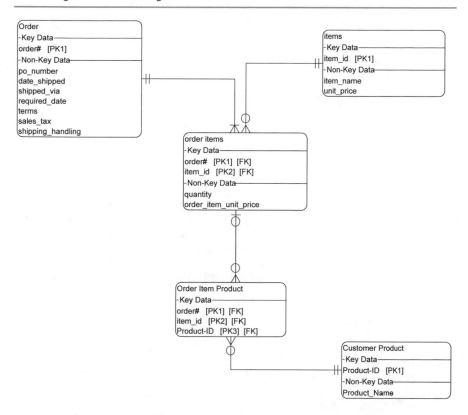

Fig. 4.23 ERD showing association between web databases and legacy employee master

them off-line. There is no desire by the Order department to create a historical tracking system. The ERD in Fig. 4.23 shows the relationships with the corporate Order Items database file.

Note that the Order Item Products entity has a one or zero relationship with the Order Item Master entity. This means that there can be an Order Item in the Order Item Product entity that does not exist in the Order item entity. Not only does this violate Normalization, it also presents a serious integrity problem. For example, if the customer wanted to display information about their products and each component Item, all Items that do not exist in the Order Item entity will display blanks, since there is no corresponding name information in the Order Item file. Obviously, this is a flaw in the database design that needs to be corrected. The remedy is to build a subsystem database that will capture all of the Order items without purging them. This would entail a system that accesses the Order Item database and merges it with the Web version of the file. The merge conversion would compare the two files and update or add new Order items without deleting the old ones. That is, the master Order Items would be searched daily to pick up new Order Items to add to the Web version. Although this is an extra step, it maintains integrity,

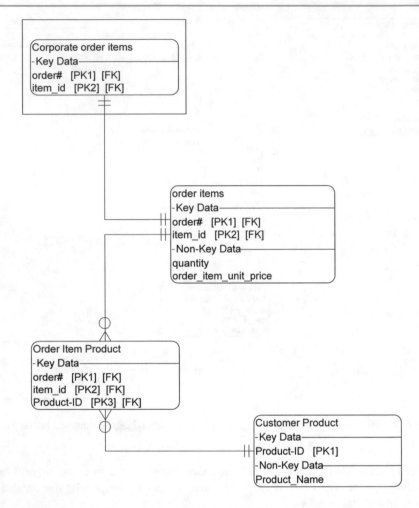

Fig. 4.24 ERD reflecting legacy link to the Order Item entity

Normalization, and most important, the requirement not to modify the original Order Item database. The drawback to this solution is that the Web version may not have up-to-date Order items information. This will depend on how often records are moved to the Web database. This can be remedied by having a replication feature, where the Web Order Item would be created at the same time as the master version. The ERD would be reconstructed as shown in Fig. 4.24.

In the above diagram the Order Item master and its relation to the Web Order Item entity are shown for informational purposes only. The master Order Item becomes more of an application requirement rather than a permanent part of the ERD. In order to "operationalize" this system, the analyst must first have to reconstruct the history data from the purged files, or simply offer the historical data as of a certain date.

4.13 Determining Domains and Triggering Operations

The growth of the relational database model has established processes for storing certain application logic at the database level. We have already defined key business rules as the vehicle to create constraints at the key attribute level. However, there are other constraints and procedures that can occur depending on the behavior of non-key attributes. Ultimately, business rules are application logic that is coded in the database language, for example PL_SQL for Oracle. These non-key attribute rules could enforce such actions as: If CITY is entered, the STATE must also be entered. This type of logic rule used to be enforced at the application level. Unfortunately, using application logic to enforce business rules is inefficient because it requires the code to be replicated in each application program. This process also limits control, in that the relational model allows users to "query" the database directly. Thus, business rules at the database level need to be written only once, and they govern all type of applications, including programs and queries.

As stated earlier, business rules are implemented at the database level via stored procedures. Stored procedures are offered by most database manufacturers, and although they are similar, they are not implemented using the same coding schemes. Therefore, moving stored procedures from one database to another is not trivial. The importance of having portable stored procedures and their relationship to partitioning databases across the Internet, Intranets, and distributed networks is becoming even more complex in mobile-based architecture. It is important to note that distributed network systems are being built under the auspices of client/server computing and may require communication among many different database vendor systems. If business rules are to be implemented at the database level, the compatibility and transportability of such code becomes a challenge. We also see that client/server will be addressed more and more as distributed and although normalization remains important, the expansion of blockchain will require multiple stored data to exist.

Business rule implementations fall into three categories: Keys, Domains and Triggers. Key business rules have already been discussed as part of the normalization process. Domains represent the constraints related to an attribute's range of values. If an attribute (key or non-key) can have a range of values from one to nine, we say that range is the domain value of the attribute. This is very important information to be included and enforced at the database level through a stored procedure for the same reasons as discussed above. The third and most powerful business rule is Triggers.

Triggers are defined as stored procedures that when activated "trigger" one or a set of other stored procedures to be executed. Triggers act on other entities, although in many database products, triggers are becoming powerful programming tools to provide significant capabilities at the database level rather than at the application level. Triggers resemble batch type files which when invoked execute a "script" or set of logical statements as shown below:

```
/* Within D. B.. only authorized users can mark    */

/* ecommerce corporation as confidential                              */


if user not in ('L','M') then

:new.corpConfidential := 'N';

 end if;

end if;

/* Ensure user has right to make a specific users private        */

if exec= 'N' then

 :new.corpexec:= 'N';

 end if;
```

This trigger was implemented in a contact management ecommerce system. The trigger is designed to allow corporate information to be marked as confidential only by specific executives. This means that an appointed executive of the corporation can enter information that is private. The second component of the trigger is programmed to automatically ensure that the executive's contacts are stored as private or confidential. These two stored procedures show how application logic executes via Oracle triggers. It is important to remember that these business rules are enforced by the database regardless of how the information is accessed.

Triggers, however, can cause problems. Because triggers can initiate activity among database files, designers must be careful that they do not impair performance. For example, suppose a trigger is written that affects 15 different database files. Should the trigger be initiated during the processing of other critical applications, it could cause significant degradation in processing, and thus affect critical production systems.

The subject of business rules is very broad yet must be specific to the actual database product to be used. Since analysts may not know which database will ultimately be used, specifications for stored procedures should be developed using the specification formats presented in Chap. 3. This is even more salient in ecommerce systems given the possibility that different databases can be used across the entire system.

4.14   De-normalization

Third NF databases can often have difficulty with performance. Specifically, significant numbers of look-up tables, which are actual 3rd NF failures, create too many index links. As a result, while we have reached the integrity needed, performance becomes an unavoidable dilemma. In fact, the more integrity, the less performance. There are a number of ways to deal with the downsides of normalized databases. One is to develop data warehouses and other off-line copies of the database. There are many bad ways to de-normalize. Indeed, any de-normalization hurts integrity. But there are two types of de-normalization that can be implemented without significantly hurting the integrity of the data.

The first type of de-normalization is to revisit 3rd NF failures to see if all of the validations are necessary. Third NF failures usually create tables that ensure that entered values are validated against a master list. For example, in Fig. 4.10, the Customers entity, created as a result of a 3rd NF failure provides a validation to all customers associated with an Order. This means that the user cannot assign any customer, but rather only those residents in the Customer entity. The screen to select a Customer would most likely use a "drop-down" menu, which would show all of the valid Customers for selection to the Order. However, there may be look-up tables that are not as critical. For example, zipcodes may or may not be validated. Whether zipcodes need to be validated depends on the use of zipcodes by the users. If they are just used to record a Customer's address, then it may not be necessary or worthwhile to have the zipcode validated. If, on the other hand, they are used for certain types of geographic analysis or mailing, then indeed, validation is necessary. This process—the process of reviewing the use and need for a validation table—should occur during the interview process. If this step is left out, then there is a high probability that too many non-key attributes will contain validation look-up entities that are unnecessary and hurt performance.

The second type of de-normalization is to add back "derived" attributes. While this is not the preferred method, it can be implemented without sacrificing integrity. This can be accomplished by creating triggers that automatically launch a stored procedure to recalculate a derived value when a dependent attribute has been altered. For example, if Amount is calculated based on Quantity * Unit-Price, then two triggers must be developed (one for Quantity and one for Unit-Price) which would recalculate Amount if either Quantity or Unit-Price were changed. While this solves the integrity issue, analysts must be cognizant over the performance conflict should the trigger be initiated during peak processing times. Therefore, there must be a balance between the trigger and when it is allowed to occur.

As stated earlier, denormalization will be occurring more often because of IoT and blockchain where portions of data will need to be distributed. I advocate for always starting the design with normalization in mind, and then depending on the network design to allow duplications based on performance of the network and the characteristics of the interface devices.

4.14.1   Summary

This chapter has provided the logical equivalent to the data component of the ecommerce system. The process of decomposing data is accomplished using LDM, which has eight major steps that need to be applied in order to functionally decompose the data. Data Flow Diagrams (DFD) are a powerful tool to use during process analysis because they provide direct input into the LDM method. Specifically, data flows provide data definitions into the Data Dictionary, which is necessary to complete LDM. Furthermore, data stores in the DFD represent the major entities, which is the first step in LDM. The output of LDM is an ERD, which represents the schematic or blueprint of the database. The ERD shows the relationships among entities and the cardinality of those relationships.

The LDM also makes provisions to develop stored procedures, which are programs developed at the database level. These procedures allow "referential integrity" to be enforced without developing application programs that operate outside the data. Stored procedures can be used to enforce key business rules, domain rules, and triggers. Triggers are batch-oriented programs that automatically execute when a particular condition has occurred at the database level, typically, when an attribute has been altered in some way.

The process of LDM also allows for the de-normalization at the logical design level. This is allowed so that analysts can avoid significant known performance problems before the physical database is completed. De-normalization should occur at the user interface time, as many of the issues will depend on the user's needs and the expansion of IoT and blockchain. Another important issue is the reduction in natural keys that are being replaced with hash algorithms to protect security.

4.15   Problems and Exercises

1. What is Logical Data Modeling trying to accomplish?
2. Define Normalization. What are the three Normal Forms?
3. What does Normalization not do?
4. What is meant by the term "derived" data element?
5. Describe the concept of combining user views. What are the political ramifications of doing this in many organizations?
6. What are Legacy Links? Describe how they can be used to enforce data integrity.
7. Name and define the three types of Business Rules.
8. Why are Stored Procedures in some ways a contradiction to the rule that data and processes need to be separated?
9. What are the disadvantages of database triggers?
10. What is meant by De-Normalization? Is this a responsibility of the analyst?

4.16 Mini-project #1

The Physician Master File from a DFD contains the following data elements:

Data Element	Description
Social Security #	Primary Key
Physician ID	Alternate Key
Last_Name	Last Name
First_Name	First Name
Mid_Init	Middle Initial
Hospital_Resident_ID	Hospital Identification
Hospital_Resident_Name	Name of Hospital
Hospital_Addr_Line1	Hospital Address
Hospital_Addr_Line2	Hospital Address
Hospital_Addr_Line3	Hospital Address
Hospital_State	Hospital's State
Hospital_City	Hospital's City
Hospital_Zip	Hospital's Zip Code
Specialty_Type	The Physician's specialty
Specialty_Name	Description of specialty
Specialty_College	College where received degree
Specialty_Degree	Degree Name
Date_Graduated	Graduation Date for specialty
DOB	Physician's Date of Birth
Year_First_Practiced	First year in practice
Year's_Pract_Exp	Practice Experience Years
Annual_Earnings	Annual Income

Assumptions:

a. A Physician can be associated with many hospitals, but must be associated with at least one.
b. A Physician can have many specialties, or have no specialty.

Assignment: Normalize to 3rd Normal Form.

4.17 Mini-project #2

The following enrollment form has been obtained from Southeast University's Computer Science program:

Student Enrollment Form

Last Name: _____

First Name: _____

Social Security Number: _____

Address Line 1:   _____

Address Line 2:   _____

City: _____

State: _____

Zipcode   _____

Course #	Section #	Course Cost
_____	_____	_____
_____	_____	_____
_____	_____	_____
_____	_____	_____
_____	_____	_____

                              Total Amount Due: _____

The students are choosing their courses from the following Course List:

Course #	Course Name	Section #	Section Time	Course Cost
QC2500	Intro to Programming	1	9:00 A.M.	800.00
		2	10:30 A.M.	800.00
		3	3:15 P.M.	800.00
		4	6:00 P.M.	800.00
QC2625	Intro to Analysis	1	11:00 P.M.	910.00
		2	4:00 P.M.	910.00
		3	5:30 P.M.	910.00
QC2790	Intro to Web Design	1	12:45 P.M.	725.00
		2	2:30 P.M.	725.00
		3	6:00 P.M.	725.00

Assignment: Using the above form, create a normalized ERD. Make sure you are in 3rd Normal Form.

HINT: You should end up with at least four entities, possibly five.

References

Date, C. J. (2000). *An introduction to database systems* (7th ed.). New York: Addison-Wesley.

Kerr, J., & Hunter, R. (1994). *Inside RAD: How to build fully functional computer systems in 90 days or less* (p. 3).

Larson, J. A., & Larson, C. L. (2000). Data models and modeling techniques. In S. Purba (Ed.), *Data management handbook* (3rd ed.). Boca Raton, FL: CRC Press LLC.

The Impact of Wireless Communication

5

5.1 The Wireless Revolution

It is important to understand how 5G wireless affects application analysis and design. In order to assess this impact, it is necessary to review 5G's technical impact on performance. In Chap. 1, I laid the foundation of how the market will likely react to increased wireless performance. In this chapter I will examine how 5G technology can be leveraged by application software developers.

Below is a summary of the expected 5G performance benefits:

1. One hundred times higher data rates than 4Gs current performance.
2. Much lower latency. A 4K video will load without any buffering time.
3. Improved handling and fixing of bandwidth issues. This will be particularly evident with new emerging technologies such as driverless cars and connected home devices.
4. Extended IoT battery life expecting to last 10 times longer attributable to the lower latency.
5. Improved communication quality in remote areas.
6. Reduced application load time by 1–2 s.

In Chap. 3, I provided an architectural approach to creating applications at the functional primitive level to expand reusability in a distributed network. The limitations of 4G latency did not allow for the required deployment to be commercially effective (Harris 2019). Edge computing will also offer much improved security which is essential for IoT deployment. Analysts will need to design rapid iterations of products that are based on real-time feedback from users. Thus, to provide timely updates, functional primitive applications will maximize the performance necessary to fulfill these requirements. Remember that functional primitives are very basic functions that come together at execution time to form more complex programs. Therefore, new use cases that can take advantage of 5G performance are:

© Springer Nature Switzerland AG 2020
A. M. Langer, *Analysis and Design of Next-Generation Software Architectures*,
https://doi.org/10.1007/978-3-030-36899-9_5

- Industrial IoT
- Cloud Augmented Reality (AR) and Virtual Reality (VR)
- Remote machinery control
- Connective automotive
- Wireless e-health
- Smart Cities.

In summary the wireless revolution is really about increased performance that changes the landscape of what we can do in a mobile environment that both increases security and lowers latency.

5.2 5G and Distributed Processing

The most impact that 5G will have on distributed processing is the decentralization of the traditional data centers. Decentralization is required to avoid bottlenecks so that the performance promised by 5G can actually be realized.

Edge computing mentioned before and discussed further in this chapter becomes the major path to figuring out how to establish a distributed network that can be monitored to improve network performance more dynamically. This is accomplished using the Edge by placing the data and program on intermediate local servers. Indeed, according to Gartner (2019), an estimated 10% of enterprise data is processed outside the central server or cloud. By 2020, they expect this to dramatically increase to 75%!

The result of these pressures on performance will result in a number of new and existing third parties competing to provide the alternative network services to their customers. Ultimately, it's about developing a strategy that allows companies to have a roadmap of alternative distributed networks where alternative paths are available depending on traffic demands. It's like running a complex railroad system where one can switch tracks depending on part failures, weather issues, and congestion. The important issue is to have as many options available for network administrators to deal with unexpected performance spikes.

It's important to recognize that predictions favor immediate investments in 5G. Indeed, by 2024 volumes of mobile data traffic are expected to increase by a factor of 5, and 25% of that traffic will be carried by 5G networks. The challenges for vendors are to provide the capacity. The hurdle for industries is to establish efficiencies through automation, artificial intelligence, and machine learning. G technology has evolved quickly—a short view of the history of G provides the exciting potential of this technology:

1G: Mobile Voice Calls
2G: Mobile Voice Calls with SMS
3G: Mobile Web Browsing
4G: Mobile Video Consumption and Higher Data Speed
5G: Technology to serve Consumers and Digitization of Industries.

5.3 Analysis and Design in a 5G World

Through the new performance improvements analysts will be able to architect a number of major new design features:

1. 3D reality as part of the user experience.
2. 3D printers will allow for 3D models of objects.
3. Evolution of chatbots will provide for real-time and prompt-based user feedback.
4. Less concern or dependency on specific hardware because cloud data centers will be doing most of the processing.
5. New life for AI and ML because of the massive quantity of data that can be collected and analyzed spontaneously.

However, the analyst also needs to consider new kinds of social experiences embedded in applications like enhanced live-streaming, and AR and VR that will improve the user interaction with the product; often from a marketing and selling perspective. That is, applications can now go beyond just doing calculations and returning values or updating data, rather adding new dimensions of experiences through photos, videos and interactive games (gamification). Many of these 3D tools will come from internal and external cloud repositories than can be included when building traditional applications. In addition, there are even more indirect affects that 5G has on other technologies:

- *Faster Web Development*: the downloading of large amounts of data simply allows for the development of more robust and content rich data and applications with much faster loading times.
- *Constant Connectivity*: reduction in power requirements which result in what is being called "Ambient Computing" which will maintain ongoing connectivity particularly with sessions that are connected to internet stations. This will allow AI and ML to track and process data near real-time.
- *Improved Augmented Reality (AR) and Virtual Reality (VR)*: will begin to become part of all user/consumer experiences. Simply put applications with 5G power will be able to contain another dimension of a user experience—integrated with voice, photos, movies as part of an application interface.
- *IoT Expansion*: more data will be expanded among sensors driving opportunities for smart cities and autonomous vehicles. This will lead the integration of BtoI, ItoI and BtoC.
- *Artificial Intelligence (AI) and Machine Learning (ML)*: speed allows applications to use AI and ML across complex distributed networks. This will make AI and ML data gathering and analysis an ongoing process.
- *Fixed Wireless*: fixed wireless is a new technology that will reduce power consumption and allow for improved interfaces that will eventually replace traditional cable and fiber optics. The result will be Wi-Fi based wireless signals direct

to customers. This will inevitably expand Wi-Fi connectivity for customers who were previously unable to get high-speed broadband.

- *Edge Computing*: dramatic increase in quasi computing centers and devices that will contain distributed databases and replicated application functional primitives.
- *Online-Video*: Video on-demand will increase and have better integration with outcome- based applications. That is, processes integrated with video experience.

By examining the above 5G indirect impact, the analyst role requires a performance criteria methodology and approach that needs to focus on latency, connectivity, and capacity. This requires additional fields to be added to the use-case format as follows:

- Required performance/response time
- Latency limits
- Associated AR and VR integration considerations
- Volume of data per interactions
- AI data gathering criteria
- ML processing needs
- Peak time load stress
- User community
- Robotic interfaces
- Other IoT device interfaces.

While various industries will provide their own performance parameters Fig. 5.1 provides the current generic key element measurements.

It is important to note that while analysts likely will not be directly engaged in setting minimum industry performance requirements, having knowledge of the technical terms and their meanings are important to effectively engage in the 5G SDLC. Many experts are now arguing that 5G is the first communication standard that will be defined for computers, that is, machine-to-machine, rather than for people!

Latency in the AirLink	< ms
Latency end-to-end (device to core)	< 10 ms
Connection density	100 x vs. current 46 LTE
Area capacity density	1 (Tbit/s/km²)
System central efficiency	10 (bit/s) Hz/cell
Peak throughput (downlink) per connection	10 Gbits/s
Energy efficiency	> 90% improvement over LTE

Fig. 5.1 Key elements of 5G Technology. Table courtesy of Mathworks

Another way to articulate 5G analysis and design is to associate it with the Six Degrees of Freedom (6DoF). 6DoF refers to, "the freedom of movement of a rigid body in three-dimensional space." This essentially means that the body can move forward/backward, up/down, left/right, and combined with rotation about three perpendicular axes (termed pitch, yaw, and roll). Figure 5.2 shows a 6DoF diagram.

From an analyst perspective this means that users can view data and applications from a myriad of dimensions and for multiple reasons. Thus, the analyst must shift from a process design approach to one that centralizes design based on repositories of data. Processes will evolve from different user views of the same data, but used for different reasons mandated by the type of outcomes desired. What is even more significant, is that the requestor may be the same physical user, but not the same logical person. This concept can be related to polymorphism in object design, where objects take on different functionality depending on its data type and class. This design challenge can be related to Fig. 1.2 which defines the different types of user identities. This means that a user of the system may be polymorphic if that person uses data as an individual and a business user, or a consumer at different sessions. Figure 5.3 shows a 5G version of 6DoF. Note that the dimensions represent the different types of behavioral views and needs of the same user but in different contexts.

Figure 5.3 represents an accelerated transformation that will require massive collections of data. Multiple functional primitive applications will dynamically be compiled to form specific applications needed by a particular user view. Many functional primitives will also be available from third party libraries. These applications are typically called functions; common routines that can be dynamically linked into more complex applications. One of the analyst's challenges is to ensure that data element attributes such as field length and data characteristics like alphanumeric, numeric, remain consistent. Depending on the user view, data

Fig. 5.2 6DoF diagram

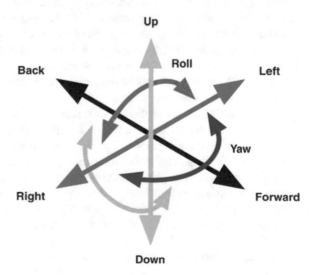

Fig. 5.3 5G version of 6DoF

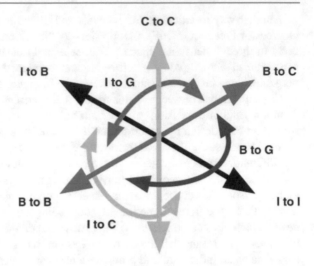

elements can take on different characteristics, so there needs to be a consensus in design to achieve unified data definitions. However, this can be problematic. In Chap. 4, I covered how data element definitions can have what is called "alias" definitions. Alias definitions allow multiple definitions of the same element, often under a different identifier. This is similar to having multiple different email addresses that are really the same personal ID.

Another factor that the analysts must address is dealing with different G systems throughout a network. It is likely that many integrated networks will have 4G subsystems because the full migration to 5G may take years to achieve. Third party vendors have developed intelligent applications that can dynamically ascertain changes in G level at the domain level. It's like software that examines your computer and determines its operating system and component features. These "connectivity management" products will assess the G network and provide an ability to execute the compatible version of an application that will work. Therefore, it will be necessary to have multiple versions of the same functional primitives! This design philosophy is certainly not unique. For example, even today many websites have alternative versions depending on weather a user is accessing the site from a computer or a smart phone.

It is probable that smartphones will eventually become the preferred 5G user device. As a result, software developers will need to maximize 5G smartphone features and functions. Therefore, we will see more smartphones equipped with more advanced capacities including more: data storage, executable applications, and interfaces with third party cloud storage. I expect that smartphones will achieve similar processing power as today's desktop and laptop computers. As one can see, the cloud interface is an important piece of 5G success. Hard disks, per se will become obsolete because the volume of data needed will make them impractical to maintain on local hard drives. Because the consumer market will drive the demand for more powerful smartphones, there will be growing pressure to make applications 5G ready. Many experts predict that 5G application development will greatly exceed the 4G evolution and occur twice as fast as 4G LTE! The transformation to 5G will, without doubt, place stress on companies to replace their slower and restricted legacy systems. Replacement of legacies then should be a high priority for any organization and could be a key for business success or failure. The paths from legacy will be discussed further in Chap. 11.

5.4 User-Generated Data and Performance Measurements

The creation of *new* data elements is another unique development that has emanated from the proliferation of social media over the internet and intertwined with the increased capabilities of AI and ML. Facebook, Uber, and Amazon have all used AI and ML to track consumer behaviors and develop statistical data elements that keep constant analysis on their behaviors. Obviously, the proliferation of sensors and other intelligent IoT devices will accelerate the collection and analysis of historical data. There are two new effects of this on data: (1) AI and ML software will be needed to do the analysis; and (2) new data elements will be created and stored in various databases across the cloud and on the Edge. These new data elements fall into the category of "derived" since they are formed as a result of a computation. A simple example is $A + B = C$. C is a derived data element (as discussed in Chap. 4) and considered redundant because if you store A and B as elements, you can always just calculate C. As explained in Chap. 4, if you store C, and A or B changes, then C is now incorrect and the database loses its integrity. However, with big data and 5G improvements, recalculations can be done much quicker. Furthermore, because these new types of derived elements are tracking changes in real time, it is necessary to add the redundancies for sake of accuracy and performance.

The Smartphone as the Key User Interface

Ericsson (2019) recently issued a report that defined what they title, "six calls to action." This report was created from a consumer survey which collected information about their preferences for interacting with cellular operations:

1. Effortless buying experience: there is considerable misalignment between what users buy and what they use. Only three in 10 smartphone users are satisfied with their plans feeling that their experiences are not simple or effortless.
2. Offer us a sense of unlimited: consumers do not want bill surprises even though 70% are not the heaviest users.
3. Treat gigabytes as currency: consumers want credit for unused gigabytes. In effect, they want credit for unused service.
4. Offer us more than just data buckets: consumers prefer plans that are personalized and based on their unique needs.
5. Give us more with 5G: 5G appeals to 76% of smartphone users and 44% are willing to pay extra. Consumers expect enhanced abilities beyond speed,

coverage, and low prices, and 50% expect enriched services. They also want single fees for 5G or fees by 5G device.

6. Keep network real for us: Consumers want the best wireless network and don't trust their operators' claims of performance capabilities.

As the smartphone becomes the gateway to most networks, it is important for organizations to select the best providers. Indeed, we already see that most of the major vendors are expanding their service offerings to include broadband services and cloud access. As they say, he/she who gets the network, will get the data—and it's all about being the provider of the data!

However, Ericsson's study exposes more about the user/consumer marketplace. It further supports the notion that all products and services must align with consumer needs. Given 6DoF any one consumer can take on many different identities. So, an individual with one smartphone may use the device in all aspects of their life, from personal use, to business applications, to consumer interfaces. I recognize that many individuals carry two smartphones to separate business and personal use; frankly there may be a better way to separate and yet integrate the two. However, according to the Ericsson study, smartphone users can be broken down into six different groups based on the types of applications and services that they use as shown in Fig. 5.4.

These six types can be mapped to Table 1.2 from Chapter 1 (shown again in Table 5.1) reconfigured in Fig. 5.5.

Ericsson's report concluded that the 5G world will make an impact quickly. Most consumers feel that 5G will be in the mainstream within 4–5 years of its launch in your region. The first phase of 5G will address broadband data needs as consumers race to buy 5G ready smartphones. Smartphones should be capable of

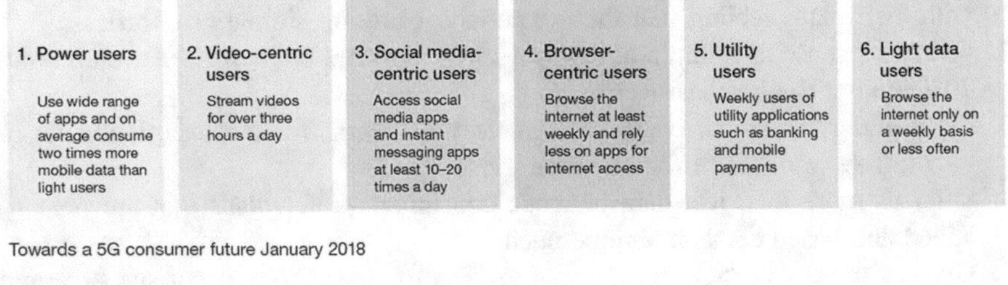

Smartphone users can be segmented into six different groups based on the types of apps and services they use on their phones and how often they use them:

1. Power users	2. Video-centric users	3. Social media-centric users	4. Browser-centric users	5. Utility users	6. Light data users
Use wide range of apps and on average consume two times more mobile data than light users	Stream videos for over three hours a day	Access social media apps and instant messaging apps at least 10–20 times a day	Browse the internet at least weekly and rely less on apps for internet access	Weekly users of utility applications such as banking and mobile payments	Browse the internet only on a weekly basis or less often

Towards a 5G consumer future January 2018

Fig. 5.4 Six user types

Table 5.1 Scope of analysis and design requirements under 5-G (from Table 1.2)

User/Consumer coverage	A&S response	Comments
Business to Business (BtoB)	Internal user and security	Current process but lacks security process
Business to Consumer (BtoC)	Internal user and external consumer and security	Current but not well integrated in most organizations
Consumer to Consumer (CtoC)	Rare except in specific trading platforms	Needs newer platforms and mobile to mobile
Business to Government	Rare except limited to information, submission of documents, and payments	Overhaul of government and business systems
Individual to Government	Rare except limited to information, submission of documents, and payments	Smart city and compliance driven
Individual to Consumer (ItoC)	Related to member portals	Limited mostly to Facebook/Linkedin
Individual to Individual (ItoI)	Knowledge based portals	Communities of practice and portals of knowledge

Fig. 5.5 Ericsson six user types and user/consumer coverage

Ericsson User Type	User/Consumer Coverage
Power User	BtoB, BtoC, BtoG,
Stream Videos	ItoC, ItoI
Social Media Users	CtoC, ItoC, ItoI
Browser Centric	BtoG, ItoG, BtoB
Utility	ItoC
Light Data	ItoC, ItoI, ItoG

downloading gigabytes of data within seconds and this is expected to be ready within 1–2 years. By the end of two years we expect earphones to provide real-time language translations, ability to watch events from multiple points using live camera streams, and aspects of virtual reality. Within five years, there should be mainstream uses of real-time augmented reality (AR), self-driving technology, connected robots, drone delivery, and 3D hologram calling to name just a few!

5.4.1   Summary

One can see that there is little time for organizations to start the processes of thinking about how such new capabilities can be implemented in their businesses, and the opportunities for new types of products and service offerings. The abundance of different versions of the same application will be stored in multiple tiers of the 5G agile application architecture that I presented in Chap. 1. The reality is how the analyst will provide the necessary requirements documentation that can define these versions and fulfill an ever changing and evolving user base.

It is evident that the challenges of understanding how to design software in the 5G era is complicated. Analysts will need to be understand a lot more of the hardware and network implications. The user community is also far more challenging as the smartphone has emerged as the gateway of interaction among a broad range of individual relationships. Performance continues to be central. With AI and ML central for competitive advantage, analyst will need to be better versed on how the placement of hardware, software and infrastructure affect the decisions on how much data to place on a device along with distributed processing in a mobile environment. I predict that we will see an emergence of Edge computing along with IoT to deliver these capabilities. Edge will also include cloud technology that will be central to the transformation of the new network era.

5.5 Problems and Exercises

1. Provide three examples of 5G advantages over 4G.
2. What is Edge Computing?
3. What are some of the effects of 5G on the analysis and design?
4. What is meant by Ericsson's six calls to action?
5. Explain what is meant by Six Degrees of Freedom and its association with 5G.
6. 5G may increase the number of new data elements. Explain.
7. What is the relationship between 5G and AI/ML?
8. What are the distinct differences in the 5G revolution as it relates to prior generations?

References

Ericsson. (2019). Six Types of Apps—Ericsson study. https://www.ericsson.com/en/trends-and-insights/consumerlab/consumer-insights/reports/six-calls-to-action.
Gartner. (2019). https://www.gartner.com/smarterwithgartner/what-edge-computing-means-for-infrastructure-and-operations-leaders/.
Harris, R. (2019). IOS Magazine. https://appdevelopermagazine.com/the-difference-5g-will-make-to-your-apps/.

The Internet of Things

6

While 5G is the initiator of possibilities to bring forth the next generation of computing, the Internet of Things (IoT) represents the actual devices that will be the vehicle for its success. For it is IoT that represents the physical components that will make a technology feasible by placing intermediate smart hardware in every place imaginable around the globe. The objective then of IoT is to allow 5G to become a reality by reducing its operational costs while increasing network reliability for consumers and businesses. The identity of IoT is to increase the uptime and real time processing of an agile architecture and to eliminate any notion of an unscheduled network failure. Thus, IoT must catch problems before they occur and provide a train that has alternative tracks should a disturbance occur during the any process. Ultimately IoT must ensure no single point of failure in any supply chain process. In order to accomplish this challenge, IoT devices must contain four components:

1. Hardware: Sensors, stress devices, friction measurements, and strain indicators.
2. Applications: Rules engine, modification of software functions, remote cooling or lubrication.
3. Analytics: AI and ML to predict failure based on assumptions and prior historical data. Handling of change capacities to avoid future failures.
4. Network: A large network or system that offers dynamic connections and alternative paths or "tracks."

IoT must accomplish five key tasks, reactive, preventive, proactive, predictive, and prescriptive. So, any analysist should start this design by ensuring that these five objectives are addressed in every process. IoT then needs multiple use cases and transforming them into a connected system that fully integrates physical devices, sensors, data extraction, secured communication, gateways, cloud servers, analytics, and real time dash boards. The following analysis and design considerations and principles must be adhered to:

© Springer Nature Switzerland AG 2020
A. M. Langer, *Analysis and Design of Next-Generation Software Architectures*,
https://doi.org/10.1007/978-3-030-36899-9_6

- Interoperability: all IoT eventually requires sensor machines, equipment and physical sites to communicate with, and the ability to exchange data.
- Information Transparency: IoT must have a continuous bridge between the physical and digital world. In other words, physical processes should be recorded and stored virtually which creates a digital twin.
- Technical Assistance: provide and display data that helps people to make better operational decisions and solve problems better and faster. IoT must especially help people to complete laborious tasks to improve productivity as well as safety.
- Decentralized Decisions: Help make decisions and execute requirements according to its defined logic.

IoT simulation is a significant part of being able to design complex IoT interface systems. Market products must provide digital prototypes to visualize how to connect devices, edge and cloud services, web, and mobile applications. All of these components make up the IoT architecture and must interact based on multiple simulation runs. IoT analytics includes dashboard and alert systems that are typically dependent on valid data sources. Consequently, IoT analysis and design must use a lean and agile approach. It must incorporate design thinking, that is, people, technology, and business all integrated in product design decisions. IoT must also be consumer centric and likely will need iterations of how business needs and consumer needs can be integrated into one requirements document. The relationship between what consumers and business need from 5G is consistent with what IoT must deliver, so the objective and relationship between these two forces must be very closely aligned.

IoT requires analysts to become more proficient with device-level application programming interfaces or APIs. These IoT APIs expose data that enables devices to transmit data to applications, so it is the data gateway. The APIs can also act as a way for an application to instruct a device, serving as a way to enact certain functions as shown in Fig. 6.1.

Figure 6.1 shows a new data element from the cloud that gets transformed by process A who sends to IoT device B. The transformed data is forwarded by device B to either a direct user or by another device/machine interface. The second example shown in Fig. 6.2 reflects the data as a value used to instruct the IoT

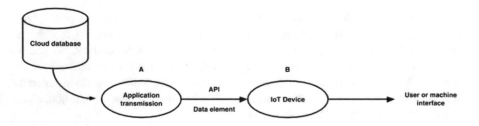

Fig. 6.1 API data transmission data flow

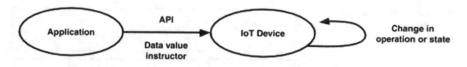

Fig. 6.2 API value transmission data flow

device B to do something based on that value—say a "1" value might mean to process a certain way.

Because so many IoT vendors might supply open source APIs, analysts will need to examine whether they need to develop the API as a requirements document or use it from a library source offered by a vendor. There is also a third possibility; use a third-party open source and modify it to work with the specific requirements of the system. In other words, there is no reason to "reinvent the wheel." There actually is nothing new about this approach in analysis. For example, functional macro libraries of routines have been available in almost every legacy architecture dating back to the mainframe. No one would think of designing a new program to calculate the square root of a number; the code exists in many libraries and it can be embedded easily in another application!

6.1   Logical Design of IoT and Communication Models

According to Mishra (2019), logical design of IoT consists of three terminologies:

1   IoT Functional Blocks
2   IoT Communication Models
3   IoT Communication APIs.

This section will map Mishra's concept to the role of the analyst.

6.1.1   IoT Functional Blocks

Functional blocks consist 6 integrated parts; Applications, Management, Services, Communication, Security, and Device as shown in Fig. 6.3.

The analyst must define each of the blocks of a device in Fig. 6.3. First, the analyst has to do the necessary use cases of each block, defining the needed data, applications, and performance requirements. As discussed earlier, many devices like sensors have software that may provide many of the services needed by a developer. However, if such APIs are not available the analyst must provide the requirements to be programmed internally. We shall see later in this book that this decision may also come under the auspices of build versus buy, especially if a

Fig. 6.3 Functional block architecture. *Source* Mishra (2019)

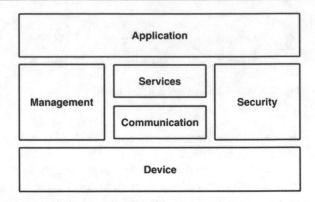

third-party API is deemed too expensive or does not contain enough of the necessary feature functions to satisfy the requirements. Such a decision may also involve determining which device best fits the needs.

6.2   IoT Communication Alternatives

There are four different types of communication alternative architectures. Multiple alternatives of course can exist across complex network systems that engage various types of IoT devices.

6.2.1   Request-Response Model

This model resembles the traditional client/server architecture discussed in Chap. 4. Although the 5G enhanced agile architecture is more distributed and less hierarchical, the client/server model is still applicable within certain device designs. As shown in Fig. 6.4 the web browser or smartphone will likely be the client in this case and the application on a device will act as the server.

It is also possible for IoT devices to integrate with multiple tiers of client/server architecture. Figure 6.5 shows an IoT devices that exists as a middle tier and providing both client and server activities depending on the requestor. In this example the dedicated server would be a cloud database likely residing on a separate physical hardware server.

6.2.2   Publish-Subscribe Model

The publish-subscribe model involves three components: Publishers, Brokers, and Consumers. A Publisher sends out data to intermediates called Brokers. Brokers

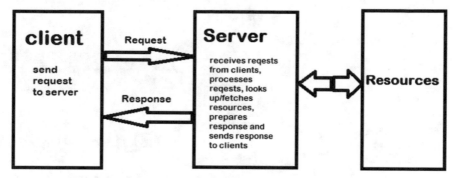

Request-Response Communication Model

Fig. 6.4 Request-response model. *Source* Mishra (2019)

Fig. 6.5 IoT Device as a Client and Server. *Source* Mishra (2019)

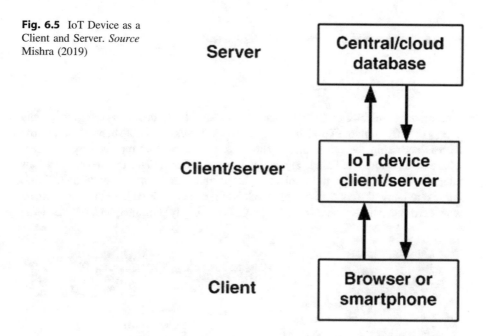

then can make the data available to a specific Consumer as the ultimate client or subscriber of the information as shown in Fig. 6.6.

Obviously, the Publisher-Subscribe model is common among data providers who work through intermediate organizations that then have some membership of users that they manage.

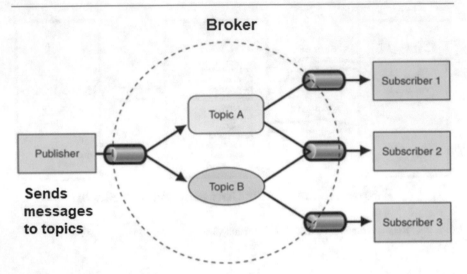

Fig. 6.6 Publisher-subscribe model. *Source* Mishra (2019)

6.2.3   Push-Pull Model

This model eliminates the intermediary broker and thus consumers are getting data access directly from the Publisher (Fig. 6.7). However, the Publisher does not know who is accessing the information. The broker is somewhat replaced by a queue, where data is stored and made available. The publisher updates the queue at various intervals. In this design the publisher has no need for a broker since there is no interest to know the consumer. This model does alleviate the dilemma of consumers who need more timely data from the publisher. Simply put, the queue defines what

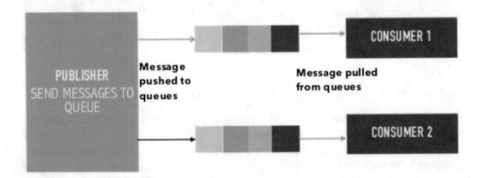

Fig. 6.7   Push-pull model. *Source* Mishra (2019)

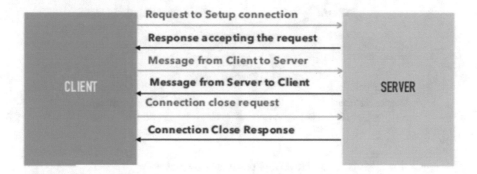

Fig. 6.8 Exclusive pair communication model. *Source* Mishra (2019)

is available to the consumer at any given time. The analyst will need to address the update frequency during the requirements gathering phase of the SDLC.

6.2.4   Exclusive Pair Model

This model is bi-directional or full-duplex with an ongoing open two-way communication between a client and server (Fig. 6.8). The server is aware of all connections from clients. The connections remain open until a client send a message to close the connection. The analyst needs to provide definitions of the messages and how the client and the server response, that is, what the messages carry and processes they enact based on the message value.

6.3   IoT as an Inversion of Traditional Analysis and Design

As previously discussed throughout this book thus far, digital transformation's major impact on analysis and design is the shift from designing systems that are product focused on performing specific user needs to one that is based more on what consumers want! Bernardi et al. (2017) states "the global economy is rapidly shifting from an economy of products to a "what if" economy" (p. 6). The authors define this shift as an "inversion paradigm" that transforms systems thinking from a product-first to a needs-first perspective. So the question that needs to be asked is, "How can technology help us reimagine and fill a need?"

While many IT professionals have historically supported this perspective, especially will agile and object design, it is the proliferation of IoT supported by 5G performance improvements that has made the world of possibilities feasible. In other words, IoT is at the center of the next wave of digital disruption and transformation as shown in Fig. 6.9.

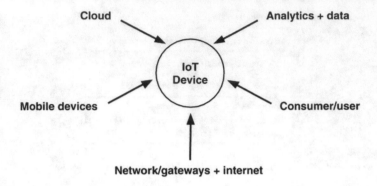

Fig. 6.9 IoT at the center

Now that I have laid the foundation and architectural models for IoT, it is critical to focus on the new roles and responsibilities of the analyst in the IoT world. This will require further transformation from a product/user perspective to one that is more functional and predictive. This is particularly evident in IoT because a device can perform many functions and cater to various consumer and machine requests. Indeed, IoT analyst must design smart objects that can integrate what is real and digital can co-exist (Bernardi et al. 2017).

6.4   Sensors, Actuators, and Computation

The IoT digital devices contain three major components and functions: sensors, actuators, and computation.

6.4.1   Sensors

Analysts must provide or identify APIs in a sensor that can measure the physical world and logically convert the information into digital data. Sensors essentially capture information, do some types of measurement, record activity and then perform applications that transform the data.

6.4.2   Actuators

Actuators actually does the reverse function of a sensor, that is, it uses digital logic on the sensor and sends messages to a physical device. An example would be to send a message to shut down a device like an oven. Analysts must define the algorithms which provide a change in a state on a device and its response to the

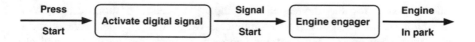

Fig. 6.10 Data flow diagram represented the computation to start a car engine

change. The response would typically be a machine-to-machine or machine-to-consumer message.

6.4.3 Computation

Computation represents the computer logic that determines the behavior between sensors and actuators. This logic results from a device sensing a situation and then applying an algorithm that instructs the sensor to send a message to the actuator to perform a function. It's like the instruction that results from pressing the ignition button on your car. The sensor receives a message and then checks the system to see if the engine in the car can safely engaged. The analyst would need to provide the data flow (Fig. 6.10) and also define the logic of the steps to take before engaging the engine. This logic likely would check the automatic transmission to ensure it is in the "Park" state before allowing the engine to be started. This algorithm would be part of the process specification of a use case.

6.5 Connectivity

The analyst must also provide definitions for IoT devices that are always connected. Such devices are always operating and initiate messages when a state changes. For example, if the temperature drops below 50% it might require the sensor to send a message to the heat device to start the heat unit. Of course, the internet provides the digital highway that allows these types of activities to occur over distances. Once again, we see how state transition diagrams can be an effective flow tool for defining logic for constant operating IoT devices.

6.6 Composability

Another interesting capability with IoT devices is their ability to communicate with each other directly thru the internet. Device to Device connectivity allows users to monitor their own systems and use a monitoring device to directly tell another device to do something. Smartphones are great examples of a monitoring device. Installing software to monitor the temperature in your house remotely for example. Composability also allows users to mix multiple monitoring communications without asking for the developer or company to make a modification. Your smartphone device can actually communicate with multiple other devices and in cases with open source can combine commands—increase temperature and also lower lights and put on music for instance.

The analysis of connectivity and composability are dependent on two types of architectures: mediation and API.

Mediation

While having devices talk to other devices seems attractive, it does have disadvantages. Having separate machine-to-machine (M2M) capabilities can cause conflict among devices and can affect the overall performance in the network. Therefore, having a "mediator" device residing in the cloud is actually a preferable solution. The mediator approach is similar to a star topology where the mediator is in the center and each device is a "spoke" as shown in Fig. 6.11.

Another benefit of a mediator is the simplification it brings when adding new devices or updating existing ones. This allows the mediator to act as a centralized hub that can coordinate all new software updates by tracking all devices across the network. The analyst may need to design a mediator or obtain a third-party product.

Suitable APIs

The design of the API is significant as it is the controlling software that holds the architecture of IoT together. The analyst should consider accumulating all the features and functions needed and then designing an inventory of APIs which

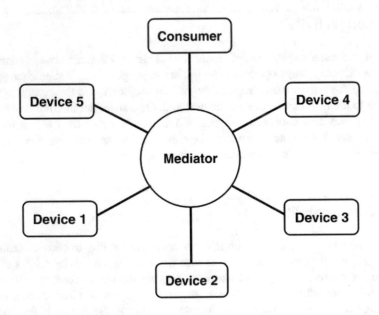

Fig. 6.11 IoT star topology using a mediator

would be stored in the mediator. The mediator, in effect, becomes the shopping outlet of all applications needed across the network. The other advantage of this design approach relates to the need to add new applications to the system. The mediator can simply be updated and used to distribute the new program to the relevant devices.

6.7 Recruitability

The idea of recruitability is closely aligned with reusability and polymorphism. Simply put it allows devices to be used for different applications. For example, a device that turns on an engine in a car, could be "recruited" by another operation that could initiate different types of operation that replicate the logic of turning on an engine. This supports functional decomposition, object-oriented analysis, and common API libraries. This also requires documentation of the functional primitive in the form of a process specification. Furthermore, the documentation should also include the routines abilities to be reused in other contexts. Finally, there are instances where a group of primitives can be combined to create functional and reusable sensors, actuators, and communication devices. This is a tradeoff decision as it establishes devices with very focused capabilities but with higher reusability.

6.8 A Word About IoT Security and Privacy

The analyst must of course be concerned with the security and privacy of any IoT device used on the network. The analyst in these circumstances must consider all the possible functions of the device and determined and its level of security exposure. I will cover much more on this topic in Chap. 9.

6.9 Immersion

Immersion is a device's ability to be shared. Indeed, devices can share its processes with other requestors if the device is available. Thus, immersion is a form of recruitment and is very useful when the original intended device is too distant in the network or is malfunctioning. The most important factor in immersion is the ability to discover available resources or devices. This can be associated with a smartphone when you "pair" it using Bluetooth technology. The analyst needs to define the context of a device's requests so that a receiving device can determine its ability to be recruited. This involves some level of intelligent messaging capability among a family of devices that can be joined under a common communication protocol.

Given the power of immersion the analyst must address several complexities:

Discoverability: not all devices can be accessed without considering certain levels of authorization. These are called static devices which essentially require both linked devices going through acceptance setup or what is typically known as a handshake. These handshakes ordinarily have two pieces, compatibility of the sharing; and identification check. Dynamic devices, on the other hand, can be linked automatically without agreement. Security in these devices, however, can also require some levels of authorization to gain access.

Context: this requires definitions of "what else" a device needs to communicate. For example, if a car device is linked to a toll booth it might also want to communicate more things about the environment, like direction, speed, license plate, time of day, date, etc.

Orchestration: this is typically a program that tracks all activity among the devices connected to the system. In many ways, orchestration can be compared to mediation in that it is a central repository of behavior among the devices.

Recruitment of Non-Digital Objects: not all objects are digital-based. As a result, non-digital objects like food need an indirect method to track and communicate with them. Typical indirect objects include RFID tagging, barcodes, and digital watermarks (that uses shades of color).

Predictive Maintenance: another unique feature of smart IoT devices is their ability to self-test and communicate conditions that require levels of maintenance. Using the network, devices can provide valuable feedback regarding their operating status so that various kinds of maintenance can be accomplished in a timely manner. These preventive maintenance capabilities all need to be part of the process specification which links hardware behavior and conditions with software intelligence.

It should be obvious that the IoT analyst must be fully equipped to deal with the myriad of details that are not only limited to traditional software design, but deeply extended to include behaviors of intelligent hardware devices.

6.10   The IoT SDLC

Many and certainly all existing organizations will need to determine how best to move forward in assimilating IoT. This will, in addition to many organizational changes, require an SDLC that addresses many of the needs discussed in this chapter. Thus, the need to modify or establish a lifecycle that managers and staff can follow.

1. Go through Object-Orientation and create functional primitives to API specifications.
2. Move or create Process Specifications.
3. Identify new and existing data elements.

4. Update (2) and (3) with any new functions based on existing and/or possible consumer experiences.
5. Design API equivalents.
6. Add available third-party APIs.
7. Map to IoT devices.
8. Select IoT configurations and interfaces based on type of device (sensor, actuator, computation).
9. Determine AI and ML functions.
10. Select type of I/O device communication functions.
11. Design communication APIs or use third-party libraries.
12. Add/modify data element definitions to be added to data dictionary including dependencies.
13. Consider connections to non-digital products using indirect connections (RFID).
14. Determine needed API and data on the Fog or Edge.

6.11   Transitioning to IoT

The way most businesses will ultimately implement IoT is through what Sinclair (2017) coined the "IoT Business Model Continuum." This model suggests that most companies cannot just overhaul their entire systems, rather they must start with their core and established business model and add the value of IoT over a period of time as shown in Fig. 6.12. The continuum eventually leads to the most important objective: business outcomes. Figure 6.12 example defines maximum business outcomes as the highest "per surgical profit" attainable. It accomplishes this maximized business outcomes by a gradual movement to efficiency by improving product-service, then service in general, to service-outcomes, before finally reaching per unit profits.

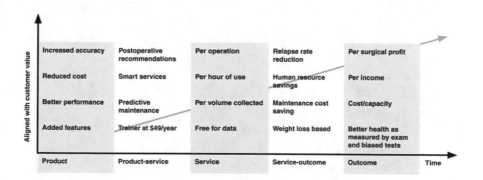

Fig. 6.12   Example of IoT Business Model Continuum. *Source* Sinclair (2017)

6.11.1 Summary

The challenge with Sinclair's model is that time may not be as available as companies think. We have experienced what digital disruption can do in an accelerated timeframe. Companies in the past that have delayed efforts to go digital have failed. The list us getting long in fact: Toys-R-Us, Nokia, and Sears certainly come to mind. CEOs and Boards must pay attention to what has occurred in the retail industry, which historically spent only 2% of their gross revenues on information technology. What is even more serious is what recently happened at GE. GE invested in a division of the company called GE-Digital which was designed to provide new types of digital services to its customers. Their initial effort has failed because the established business units used the new digital division to support its legacy needs. Bottom line GE never generated the forecasted revenues from new business. The message here is to be careful with the existing and dominant core businesses; it appears historically such units tend to unconsciously do everything they can to preserve the old way of doing things! Remember that many people feel that the majority of IoT products will be developed without a plan!

Next we need to understand Blockchain analysis and design and its contribution to security!

References

Bernardi, L., Sarma, S., & Traub, K. R. (2017). *The inversion factor: How to thrive in the IoT economy*. Cambridge, MA: MIT Press.

Mishra, H. (2019). Logical Design of IoT | IoT Communication Models & APIs. https://iotbyhvm. ooo/logical-design-of-iot/.

Sinclair, B. (2017). *IoT Inc: How your company can use the Internet of Things to win in the outcome economy*. New York: McGraw Hill Education.

Blockchain Analysis and Design

7

7.1 Understanding Blockchain Architecture

Blockchain technology represents an interesting architectural invention that will primarily addresses the challenges of cyber security in the internet. As I previous discussed, the existing central database architectures cannot provide the security necessary to launch IoT systems. Blockchain is defined as a "ledger-based" system, in that it is designed to track all transactions and update all members of the chain. In actuality, the block chain design evolved from what started as a linked list data structure. Essentially, a linked-list originally was designed as a data structure that linked to another data element by storing information about where a value was stored was stored in memory. It is a pointer system, that shows prior (previous) links and forward links as shown in Fig. 7.1.

The problem addressed by linked list data structures was its ability to store related data or values without requiring the physical storage to be sequential. In other words, by having these "links" related file elements could actually be stored in different parts of storage and on different physical devices as shown in Fig. 7.2.

The importance of the linked list data structure is that it allowed a logical file of information to be stored physically in different locations. But to the user it was invisible and allowed the system to maximize the data storage. However, common with any data structure strategy, there is always a downside. The allocation of a logical file across multiple physical devices diminishes performance. With large logical files that are distributed across many devices or even fragmented on the same disk will have significant reductions in performance.

Blockchain evolved the linked list model for similar yet different objectives. The linked data structures shown in Fig. 7.1 are now more complex and defined as "blocks" of information. The architecture allows for new blocks to be added dynamically and simultaneously updates each block when changes are made. A block, rather than representing a data element actually represents an account or

A. M. Langer, *Analysis and Design of Next-Generation Software Architectures*,
https://doi.org/10.1007/978-3-030-36899-9_7

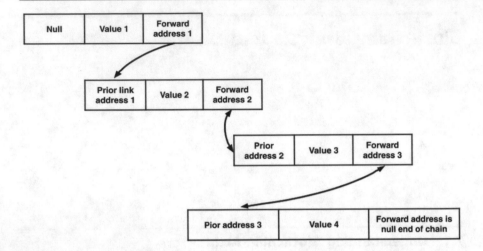

Fig. 7.1 Linked list data structure

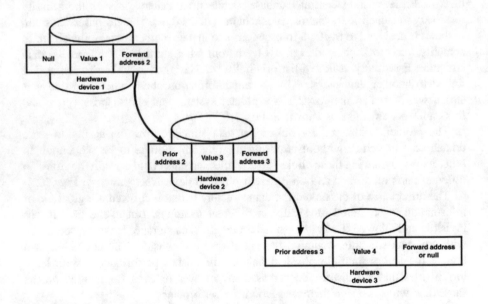

Fig. 7.2 Linked list storage across physical storage devices

user. So, each individual has their own block in the chain. Depending on the design of the blockchain (or the blockchain product) users may have equal or unequal rights on how they access and update the other blocks in the chain. All blocks contain information like the date, time, and amount of a transaction. In effect, the blockchain architecture acts as a modified linked list designed to keep track of transactions as opposed to link data elements. For this reason, this blockchain is a perfect solution for house-keeping and is often referred to as a ledger-based technology.

The word ledger really emanated from the accounting profession, where ledgers were created to keep track of detailed transactions often known as debits and credits. At any given time, the accountant need only add and subtract all the transaction amount entries to calculate the balance of any ledger account. The important feature of a ledger; however, is its audit-trail feature that ensures knowledge of every transaction that makes up the balance. It acts like a running total at all times. What is also essential to a ledger is its ability to recalculate the balance each time so that a given balance can be tested for accuracy. Furthermore, the source date of every transaction in a ledger is documented. In blockchain, the ledger is the account; every account has a unique ledger in the chain. Another important factor in any ledger is the modification of any transaction is not allowed. For example, when you need to adjust an accounting entry, you don't directly modify the original transaction or entry, rather one enters an "adjusting" entry that modifies the balance. Thus, in a ledger you can only create a transaction or read it, but you cannot modify or delete! Blockchain follows this rule and this is why it provides two significant benefits, (1) complete audit trail of all behavior in the block, and (2) you cannot reverse or modify a transaction which really restricts hackers. So, every blockchain entry is documented and a block stores the associated authorization and date/time of each transaction.

In the blockchain, a user ledger is identified through a "hash" code key. The hash code is a random-based calculated number that is extremely difficult to decipher, so it adds a strong security to the chain. Each member of the blockchain has access to all the blocks and keeps a separate copy of the blockchain on their resident network system. This means that when a block is updated or a new block is added, each copy must be updated, which of course brings back the challenge of latency in performance. However, because there is no central controlling copy of the blockchain, it makes it challenging for a hacker to manipulate every copy. This is where blockchain adds hope to establish an internet architecture that can protect users.

The blockchain protocol is built on the concept of "consensus." Due to latency issues, the consensus protocol will always assume that the longest chain in a blockchain version represents the one that users trust most. So, in complex and large blockchain's where updating is constantly occurring, the longest typically is the one that is most current at any instance. Of course, the importance of consensus is directly related to the size and the amounts of new blocks coming into the chain. Blockchains are also both private and public. From a public perspective anyone can view the contents of the blockchain but cannot access accounts without a private key which allows a user to transfer an item to the block. Another important factor to note is that the blockchain infrastructure provides trust (better known as access to the block) but not trust between blocks or users directly as this maximizes the security benefit of blockchain. There are various types of trust which means, how do I know you are the real authorized person accessing the block. There are six common proofs that are implemented in any blockchain product:

1. *Proof of Work*: avoids a hacker because the network machines are required to prove which involves complex algorithms not available to hackers. Further, the network machines must have specific configurations and space to be able to complete the algorithm. Proof of Work tends to be the most attractive to developers.
2. *Proof of Stake*: requires the user to prove ownership of a specific amount of money. This approach is more common in Bitcoin blockchain given the trading of cryptocurrency. Simply put the owner must prove they have the money they are attempting to trade.
3. *Proof of Hold*: the user has more rights based on the amount of time of possession of the coin.
4. *Delegated Proof of Stake (DPOS)*: this allows users known as delegates who want to produce new blocks on the network. Delegates are allocated blocks based on the highest number of votes they receive from other delegates. DPOS is relevant when access to the blockchain is valued by having multiple blocks or accounts.
5. *Proof of Capacity*: this is an algorithm that requires users to solve challenges in the form of a puzzle. The more storage the requestor has allows them to solve the puzzle quicker. The puzzle is created by the service provider.
6. *Proof of Elapsed Time*: users are randomly assigned a wait time. Those that have shorter wait times get access first.

As one can see these proofs for access are very effective on keeping out hackers as opposed to whether who have rights as a user. This is why blockchain is so attractive as a solution particularly for IoT applications.

7.2 Forecasted Growth of Blockchain

The growth of blockchain technology is predicted to grow enormously. Deloitte in 2019 recently surveyed 1,000 companies in seven countries and found that 34% already had blockchain in production with another 40% planning to invest 5 million in the year 2020! Much of this explosion of interest can be attributed to three factors:

1. The launch of Bitcoin in 2009 was the first successful blockchain implementation;
2. The coming of 5G that addresses the latency criticisms of blockchain architecture; and
3. The need for a cyber security infrastructure that allows IoT to be protected.

There are certainly particular industries that are earlier adopters of blockchain, specifically banks, healthcare, property record keeping, smart contracts, and voting to name just a few.

7.2.1   Advantages and Disadvantages of Blockchain

Although blockchain represents a decentralized solution that favors IoT devices, there are specific advantages and disadvantages to consider:

Positives:

- Accuracy in verification
- Elimination of 3rd party verifiers lowering cost
- Security via decentralization
- Transparency
- IoT and 5G capabilities
- Scalability
- Auditability and traceability
- Better access to data

Negatives:

- Increased costs for technology
- Latency and performance issues still exist
- Attraction to hackers
- Short history of results

As one can see above will there are disadvantages, there appears to be more positives and hope that blockchain can evolve into the next generation of a new architectural design.

7.3   Analysis and Design of Blockchain

Many of the blockchain decisions have been focused in two areas; (1) the requirements analysis to determine the feasibility of a blockchain implementation, and (2) the architectural infrastructure decisions on the rules and governance of the blockchain itself. While I will provide various examples of these two issues, the important objective of this section is to define what responsibilities an analyst has in this process as it relates to the SDLC.

Table 7.1 Generic blockchain business process requirement definitions

Generic requirement	Description
Data storage	In block/out of block
Mode location	Where are spokes in the network mobile, IoT etc.
Network bandwidth	G power (5G, 4G etc.)
Type of blockchain	Public, private, hybrid, consortium
Industry	
Customer experience factors	User friendliness, robustness, accessibility
Overall goal of the system	Paragraph of objectives
Actors	Human and machine to machine
Authority level	Trusted, decentralized
External system interlaces	Other networks
Data structures	External interfaces
Internal functions	Modifiers
Tests	Security assessment
External subsystem	User stories, inserting new blocks, acceptance tests of the system, user interface
Precondition	Requirements to become an actor

Overall, when assessing blockchain use cases are the preferred method for setting up and selecting the right blockchain design. Use cases for blockchain must first focus on specific features and functions that are common to an industry. Along with that specific industry comes possible processing regulations as well as technical requirements including such things like smart contracts, cryptocurrencies and legal constraints. All of these must be part of the analysis function. Further, data requirements and speed of response times are critical technical issues that drive the feasibility of a blockchain solution and its selection. So, it's not just feasibility, it's the overall construct of the blockchain itself, although in the financial industry one can see why block chain, while attractive, is of concern when it comes to the latency issue of performance capacity and scalability. In addition, the analyst must address the size of transactions and the storage needs within a block in the chain. Table 7.1 provides a list of the generic types of things that analysts should be prepared to capture and document.

Essentially blockchain in analysis and design resembles the licensing of a transaction processor. It's almost like selecting a type of engine for a car. Engines have varying capabilities and limitations depending on the type of car you wish to own and how you want to drive the car. Things like gas mileage, pickup, durability, dependability, and the way the engine shifts, etc. In order to install the blockchain engine you must also understand the best way to install the product. Installation again would need to reflect industry requirements, performance preferences and regulatory limitations. The settings needed are part of the selection considerations for the ultimate blockchain vendor and product you choose to license. Each vendor, as with all third-party vendors have advantages and disadvantages in the product they license. Of course, each vendor believes they offer the best product, but there are also blockchains that might be better for a particular industry like Bitcoin in the financial markets.

There exist hundreds of industry and technology use cases prototypes for that can assist analysts to install and set up the blockchain once the product has been selected. It might even be part of driving the decision; let's say the vendor did not support a feature that is required buy your industry. An example of a roadmap for design is Yrjola's (2019) use case for a citizens Broadband Radio Service Spectrum. His model also provides a flow diagram of interfaces, a life cycle flow of a transaction, and a decision diagram for setting up the blockchain as shown in Fig. 7.3.

Another suggested approach to determining needs is to design process flows— sort of a data flow/flowchart of the sequence of events that occur when a user gets on the system and the decisions trees that define the flow. Xu et al. (2017) provides a sample design process that demonstrates how analysts can document requirements and the logic flow in Fig. 7.4. inevitably logic flows also uncover data element needs.

Blockchain design can also make use of traditional analysis tools covered earlier in this book. Various agile analysis tools can be incorporated in creating a design document for a blockchain smart contract system created by Marchesi et al. (2018). Their approach can be used as a guideline for a generic blockchain SDLC. In their particular solution the authors use the UML methodology, but analysts can use they choose any of the structured analysis approaches.

Step 1: State the goal of the system in the form of a few paragraphs.
Step 2: Identify the actors (both human and machine) that ultimately represent the boundaries of inputs and outputs.
Step 3: Create a development process diagram as shown in Fig. 7.5.
Step 4: Develop the system requirements using high-level user stories using both prose writing and a use case diagram as shown in Fig. 7.6.
Step 5: Translate the user stories into object-class diagrams as shown in Fig. 7.7 which shows the entities, data structures, and operations.

Use case	Shared write	Absence of trust	Disintermediation	Interaction	Confidentiality
SAS-SAS data exchange	+	+	+	+	Hybrid
SAS marketplace	+	+	+	+	Hybrid
Sensing as a service	+	+	+	+	Hybrid
Element tracking	+	+	+	+	Hybrid
Neutral hosting	+	+	+	•	Hybrid
Operator roaming	+	+	+	+	Hybrid
CBSD measurements	+	-	+/-	-	Private
FCC database	-	-	-	-	Private
ESC sensing	+	+	-	-	Private

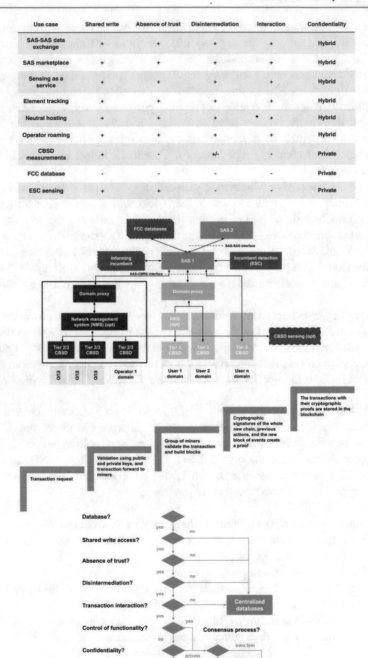

Fig. 7.3 Yrjola Broadband Radio Service Spectrum blockchain

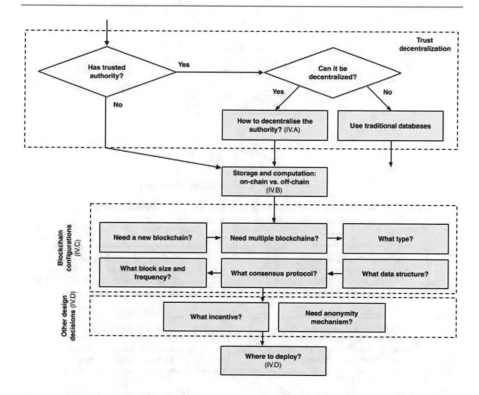

Fig. 7.4 Xu et al. design process for blockchain-based systems

Step 6: Develop State Transition Diagram showing possible states and what flows can cause a change in state. The example in Fig. 7.8 reflects UML style state chart. Step 7: Create process specifications of functions from User Stories as shown in Table 7.2.

The Marchesi et al. sample depicts how requirements documents might appear in the SDLC for a typical smart contract blockchain. While there can be many variances, it is important to recognize how traditional analysis tools can be embedded in the block chain engine that will be inserted into the various other processes and data interfaces of the larger applications needed in any complex system.

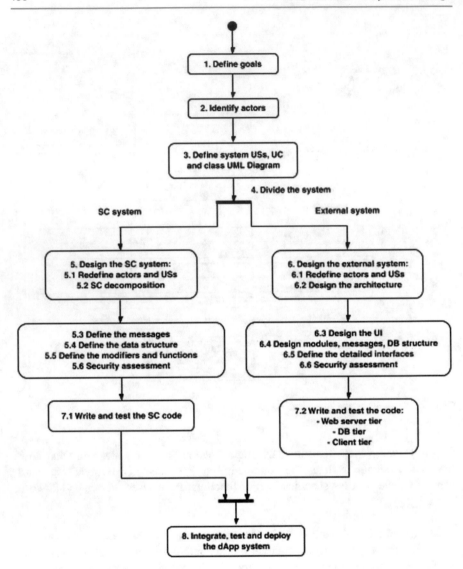

Fig. 7.5 Sample development process for a smart contract system

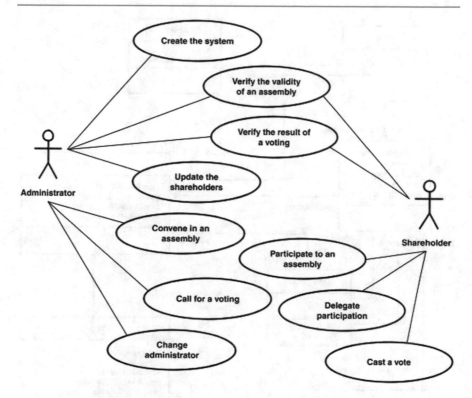

Fig. 7.6 User story flow diagram

7.4 Summary

Blockchain represents an essential architectural component to make IoT a feasible and secure engine that can be incorporated into complex systems. Essentially the blockchain will serve as the validation and recording of transaction using an accounting ledger-based system that guards against hackers. Without this engine system IoT cannot achieve widespread use across various industries and specific technologies. While we are still in the embryonic stages of blockchain development and it is predicted that there will be a plethora of third-party blockchain products that will provide the architecture for specific industry-related product processing.

However, it is important to recognize that the latency issue is still very much a challenge. The philosophy of the blockchain architecture in terms of its update process very much resembles the old IBM token-ring structure. As a recollection of that network design, IBMs computer nodes each had to be updated in a circular ring that got updated sequentially. The problem then was that the network design was too slow to achieve itself as a feasible solution to networked personal computers.

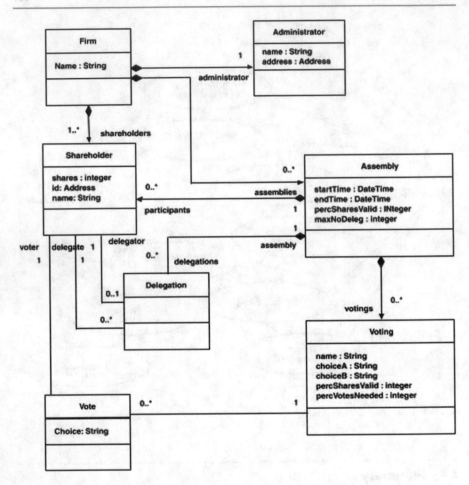

Fig. 7.7 Object-class diagrams derived from user stories

Figure 7.9 depicts the IBM token ring structure. The blockchain architecture replaces the personal computers with blocks. While updating blocks is faster than token ring network structure, it still will be challenged in scaling, especially in a mobile network system. While 5G will make blockchain more feasible, scalability across large mobile networks may still hinder widespread use. The next chapter will address some of the potential solutions that can ultimately increase processing power to support more scalable blockchain development.

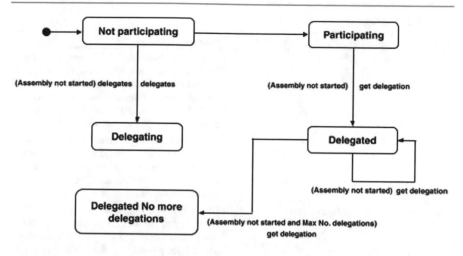

Fig. 7.8 State diagram derived from object-class flow

Table 7.2 Process specifications of functions from user stories

Function	Modifiers, parameters	Action—Notes
constructor	string nameFirm string nameAdmin [(string nameSh, address addrSh, unit 16 noShares)]	Create the Voting Management contract, inputting the name of the firm, the Administrator's name and, for each shareholder: name, address and number of shares. Add a new shareholder, giving his name, address and number of shares
addShareholder	onlyOwner string nameSh address addrSh unit16 noShares	Add a new shareholder, giving his name, address and number of shares
Delete Sharehold	onlyOwner address addrSh	Delete the given shareholder, giving his address. Can be done only it the shareholder has no active participation in an assembly
editShareholder	onlyOwner address addrSh string nameSh unit 16 noShares	Update the given shareholder, giving his address (that cannot be changed), name and number of shares. Can be done only if the shareholder has no active participation in an assembly
Change Administrator	onlyOwner address newOwner string nameAdmin	Give the address and the name of the new administrator
Convene Assembly	onlyOwner	Convene an assembly, giving start and end date and time of the assembly, a short description, the minimum percentage of shares needed for its validity, and the maximum number of delegations that can be given to a single Shareholder. No existing assembly can overlap with the new one

(continued)

Table 7.2 (continued)

Function	Modifiers, parameters	Action—Notes
addVoting	onlyOwner	Add a call for voting to the given assembly, specifying the name of the voting, the two options that should be chosen, the minimum percentage of voting shares, and of votes needed to have a valid vote The assembly must not have already started
participate	onlyShareholder	Register the participation of the sender to the given Assembly, provided that the start date and time of the Assembly has not yet passed, and that the sender has not already delegated another Shareholder, or already registered
delegate	onlyShareholder	Delegate the participation to a given Assembly to another Shareholder, provided that the start date and time of the Assembly has not yet passed, that the sender has not already registered his participation of delegated another Shareholder, that the delegated Shareholder has registered to the Assembly, and has not yet reached the maximum number of delegations
castVote	onlyShareholder	Cast a vote for one of the choices of a given voting, provided that the sender is participating to the Assembly of the voting, that this Assembly has started and has not yet expired, and that the vote has not already cast
verifyValidity *view*	OnlyOwnerOrShareholder	Read the total number of shares that participated to a given Assembly, and check if the minimum number has been reached. The Assembly must have expired
readResults *view*	OnlyOwnerOrShareholder	Read the voting results (choice 1, choice 2 or no choice), given an Assembly, and the name of a voting. The Assembly must have expired
deleteContract	onlyOwner	Permanently delete the contract

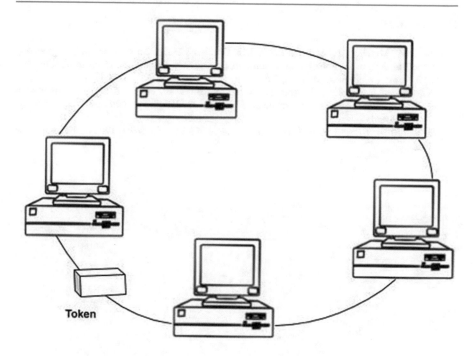

Fig. 7.9 IBM Token Ring Architecture

7.5   Problems and Exercises

1. Explain the relationship between Blockchain and Linked Lists.
2. What is meant by a hash code?
3. Compare the positives and negatives of blockchain.
4. What are the two main objectives of blockchain analysis and design?
5. How does use cases relate to blockchain analysis?
6. What is the relationship between a use case diagram, prose writing, and a minispec?
7. What is the relationship between a minispec and pseudocode?
8. How can user stories be used to complete the requirements of the blockchain modeling tools?
9. What is the relationship between a state transition diagram and blockchain? Why is it important?
10. Why is blockchain architecture so important in the design of mobile-based IoT systems?

References

Marchesi, M., Marchesi, L., & Tonelli, R. (2018, October 12–13). An agile software engineering method to design blockchain applications. In *Preliminary Version Accepted at the Software Engineering Conference*. Moscow, Russia.

Xu, X., Weber, I., Staples, M., Zhu, L., Bosch, J., Bass, L., et al. (2017). A taxonomy of blockchain-based systems for architecture design. In *IEEE International Conference on Software Architecture*.

Yrjola, S. (2019). *Citizens Broadband Radio Service: WinnComm 2019*. Technical Presentation Abstracts.

Quantum Computing, AI, ML, and the Cloud

8

As discussed in Chap. 1, quantum computing while not yet scalable has the potential to change the processing capabilities of computing especially for ML and AI processing. Without getting into the detailed hardware technicalities, the essential advantage of quantum is that it can evaluate many potential answers simultaneously (superposition) that results in improving calculation speeds immensely. Traditional computers behave in a sequential manner, but quantum allows multiple calculations to take place simultaneously and yet be related to the same problem. It's like having dimensions of processing but somehow offering one solution. Because of this advantage, there are also disadvantages. Specifically, quantum provides value for certain types of computational problems. Where such computational algorithms are not to the advantage of quantum there is no performance improvement from traditional binary-based computers. The true benefit then of quantum computing is dealing with uncertainty problems These are also known as "quantum algorithms" which can solve difficult equations in many different ways. For example, the successful quantum algorithm called factorization created by Peter Shore of AT&T Bell Laboratories proved that quantum could factor large numbers into their prime factors in seconds compared to a classical computer that could almost take forever. These types of benefits tend to favor performance improvements in ML and AI which require large data crunching to solve or analyze complex datasets. One can see then that quantum is a very attractive alternative to speeding up computations that can provide incredible results for predictive analytic issues. Picture the value that quantum brings when analyzing causes of disease, or maximizing optimizations across sectors and use cases for smart cities, traffic systems, lights, meters, utilities, buildings all at the same time. Quantum allows for these simultaneous computations to take place and yet maintain a relationship with each other (called entanglement). Traditional computers would need to analyze each computation sequentially ultimately limiting scalability.

© Springer Nature Switzerland AG 2020
A. M. Langer, *Analysis and Design of Next-Generation Software Architectures*,
https://doi.org/10.1007/978-3-030-36899-9_8

8.1 Datasets

With enormous processing speed potential, quantum computing allows for quicker analysis and better integration of distributed large datasets of information. This is accomplished using extensive search and determinations of the patterns that exist in data that otherwise could not be accomplished to have an impact on business applications. Furthermore, quantum computers expand the potential to examine large databases that could be distributed across multiple network and machine platforms. Finally, datasets, databases, and other data structures can all be investigated to render correlations that provide valuable probabilities. As quantum evolves it can dramatically alter hardware architectures, accelerate the proliferation of IoT devices that may indeed result in changing the way companies use their data for competitive advantage.

8.2 IoT and Quantum

Economist's Business Insider Intelligence (2018) has forecasted that by 2023 consumers, companies, and governments will install 40 billion IoT devices globally. The result of these installations will generate massive data on a daily basis. The challenge for any analyst will be to understand how to approach the processing of this data to ultimately generate useful information and knowledge. I have already established the importance of IoT security. Think further of this importance as these IoT devices begin to generate what will likely be very sensitive information. Therefore, for IoT to reach its potential the confidentiality of consumer data must be protected and even guaranteed. Another interesting advantage of a quantum machine is its potential to generate secured systems using cryptography from quantum fed algorithms that require very large machines. These machines, as I have previously mentioned are typically not available to hackers. In theory, however, quantum cryptography should be able to generate keys that are totally random, unique, and incapable of being replicated. And with the speed of quantum calculations, a unique long key can be generated per transaction.

8.3 AI, ML and Predictive Analytics

Having established that quantum's role is to help crunch massive amounts of data, the next challenge is to determine how to collect the data, store it, and what algorithms will be needed to obtain valuable information that can be used to make predictions. One must accept the fact that the voluminous data certainly is beyond the human capacity to derive meaningful predictive data, not to mention how long it might take to even if an individual could process the data!

In the past, obtaining data and analyzing it to make predictions required trained personnel in areas such as mathematics, statistics, and computer science. However, today there are now advanced APIs that can be obtained that can allow non-technical people to get results. So, again the analyst needs to start thinking of the data especially identifying what that data does and where it should reside!

The strategy relating to predictive analytics is quickly becoming more focused on the ML component of AI. The reason for this development is simple: most organizations likely do not know or understand the data they possess. Yes, they might know the business elements they store, but ML provides a whole new opportunity. For example, most business environments are broken down into functional units or divisions. Individuals that operate within these divisions are often siloed in such a way that they know their own data, but little about the data in other divisions that contain important related data that can provide value not only for their division, but the business overall. Thus, they do not know what to search for, because they know little of the opportunity. Furthermore, there is massive information that tends not to get stored in a way that allows for easy understanding of value. Transaction data updates databases which in turn is used to gather the information for reporting and analysis. This is particularly true for different types of transactional data that comes from consumers as well as data that results from their behaviors. Anyone who has spent time on Amazon or other consumer sites has experienced the application offering the consumer other opportunities based on their search behavior on the retail website. These behaviors are stored in the form of transactions and then fed into datasets that can be analyzed by machine-driven algorithms.

With any new software opportunities comes dangers, and such is true with ML. Below are some of the setbacks that can occur:

1. Lack of transactions or examples that can render dependable and generalizable results to draw conclusions.
2. Similar inputs can sometimes render different outputs. To have effective predictions there must be a clear relationship between inputs and outputs.
3. Mistakes in categorization of data is a killer. We have all seen the problems when a data element is not defined appropriately.
4. Incorrect or inappropriate example used. These are cases where certain factors are not considered. We see this occur when certain factors affect consumer habits that were not considered in the use case.
5. Challenge of tagging data and then classifying all of its relationships.
6. Implementing Natural language Processing where inputs are purely textual and outputs are often categorical. This challenge is how to take textual input and determine an output value such as positive or negative.

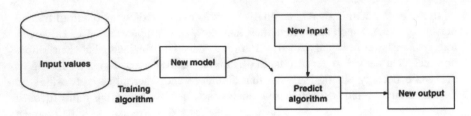

Fig. 8.1 Example of training to prediction algorithms

There are two types of ML algorithms:

1. Training: learning a model from examples. This algorithm is also known as a learning algorithm because it examines sets of inputs and outputs and crates a new model based on the datasets.
2. Prediction: takes an existing model with a new input and returns an output value as shown in Fig. 8.1.

What occurs in Fig. 8.1 is a simple way of predicting a value that might be missing from a dataset. This can often occur from converted data from a legacy or earlier version of application data. By examining the new application data ML and establish a model that might predict what that missing element would have been in the old system as shown in Fig. 8.2.

Figure 8.2 shows that the legacy system did not store or capture a student's graduation rate. In the new or replacement system, a data element was added to capture the student's graduation rate. When the legacy records are converted ML could use a training algorithm to review the relationship between GPA Score and Graduation Rate to see if there is a correlation among the records in the new system. Should that correlation have predictability and the dataset in the new system is large

Legacy on old database record

Last name	First name	GPA score

New or replacement database record

Last name	First name	GPA score	Graduation rate

Fig. 8.2 Updating legacy data element using ML training and prediction algorithms

enough, then the prediction algorithm could derive a graduation rate from the student's GPA Score in the old system and calculate a Graduation Rate. We can see from this example that training algorithm can indirectly be used to create the prediction module in a sequence. The size of the dataset that would likely be deemed appropriate based on forms of statistics theory.

8.4 ML in a Service Environment

ML can be designed as a service in the cloud. In this design an ML program and datasets can be stored on a separate and powerful server (preferably a quantum processor!) that could provide the necessary performance needed to deliver results quickly. This type of network architecture can be developed using three types of ML APIs according to Dorard (2014) as follows:

1. *Specialized Prediction*: these APIs do very specific tasks like determining the speaking language embedded in a text. Specialized prediction APIs may often be available from third party libraries. Because these APIs specific but common they tend to be easier to implement.
2. *Generic Predictor*: this API represents the training to predictor algorithm example shown in Fig. 8.2. So, there are two algorithms needed in the generic predictor, one to create a training model based on previous data and one that utilizes the training model to deal with new input. Generic predictor APIs are particularly effective for regression problems (algorithms that predict a real value).
3. *Algorithm APIs*: these APIs, while similar to Generic Predictors are much more focused on a specific problem, so the parameter must be very specific. Indeed, think of algorithm APIs are specialty problem solvers. Should a specialize algorithm API not be available, then a generic API can be used by adding training data.

8.5 Analysis ML Use Cases

An analyst can participate in ML design by providing types of use cases for developers. Dorard (2014) provides a format to follow:

- WHO: who does the example concern?
- DESCRIPTION: what is the context, and what are we trying to do?
- QUESTIONS ASKED: how would you write the questions that the predictive model should give answers to in plain English?
- TYPE OF ML PROBLEM: classification or regression?
- INPUT: what are we doing predictions on?

- FEATURES: which aspects of the inputs are we considering, and what kind of information do we have in their representation?
- OUTPUT: what does the predictive model return?
- DATA COLLECTION: how are example input-output pairs obtained to train the predictive model?
- HOW PREDICTIONS ARE USED: when are predictions being made, and what do we do once we have them?

As one can see the analysts provides a guideline of questions to be answered as opposed to the answers to these questions. Clearly ML design then requires subject matter experts to answer these questions, or a consumer/user community that can define certainly the output needs.

Below and using Dorard's Pricing Optimization example shows the use case with actual answers to the questions:

- WHO: Shops, stores, and sellers.
- DESCRIPTION: We are introducing a new product within existing category of products that are already being sold, and we want to predict how we should price this new product; the product could be, for example, a bottle of wine in a wine shop. Or a new house for sales.
- QUESTIONS ASKED: "What should be the price of this new product in this given (and fixed) category?"
- TYPE OF ML PROBLEM: Regression.
- INPUT: Product.
- FEATURES: Information about the product, specific to its category. In the wine bottle example, this could be the region or origin, the type of grapes, or the rating from a wine magazine. In the house example, this can be the number of bedrooms, bathrooms, the surface, the year it was built, or the type of house. We can also include a text description, and, when relevant, the cost to manufacture the product and the number of sales (total or per period of time).
- OUTPUT: Price.
- DATA COLLECTION: Every time a product of the same category was sold, we log the price at which it went. Note that the same product might be sold several times (or not) and at the same or different prices, which affect the number of training data points.
- HOW PREDICTIONS ARE USED: We set the price of the product to the value given the predictive model (no need to add a margin, this is already incorporated by the nature of the training data). Note that if the number of sales is one of the features, we need to do a manual estimation of this for the new product before we can make a prediction on it. Besides, since prices are likely to change over time, it is important to frequently update the predictive model with new data.

8.6 Data Preparation

One can see that the most important aspect of ML is the quality of the data. To no one's surprise it's also the most challenging problem to solve in business. Many traditional companies likely have proliferations of data across multiple systems from the start of business computing in the early 1960s. Notwithstanding what companies have compiled in central systems, the amount of local data stored across local area network systems from the 1980s is significant. There is also a plethora of databases stored on PCs from desktop products like Excel, Foxpro, and Access to name just a few. In addition, there is rich data that is stored in text-based files. While the challenge seems overwhelming, progress has been made with the development of sophisticated natural language products that can specifically extract useful data from unformatted data. My point; however, is that analysts need to shift their focus more on analysis of data than on process. With the proliferation of IoT the issue of cleansing data for ML is more important than process analysis. Do not misunderstand my point. I am not advocating that process analysis is no longer necessary or important, rather that data quality need more attention than before. So, from an analyst perspective the data process should focus on the following steps:

1. Locate where data resides across the enterprise.
2. Understand the differing formats of the datasets and/or type of files systems where the data is stored.
3. Determine the meaning of each data element that comprises a file record.
4. Identify text-based files and see if natural language processes can aid in defining the data need by ML algorithms.
5. Extract the data elements that are needed for ML from various datasets.
6. Do a quality review based on the results of test runs on the extracted data.
7. Automate the extract programs and implement an ML API.

Another aspect of data extract is to decide whether to place the data in a central repository cloud system. This objective always sounds doable on paper, but turns out to be an overwhelming task and often fails to achieve its objectives for a number of reasons. So, for now the all-powerful central database will be left for discussion in later chapters. The argument to merge everything has far more benefits and disadvantages beyond creating ML solutions.

8.7 Cloud

With the advent of 5G, IoT, blockchain, and potentially quantum, the cloud is assured to a critical part of the quest to obtain better speed, centralization, and security in a digital world. The challenge is how best to design the cloud architecture, that is, whether to have a private or public cloud or some combination. Further, once the infrastructure is designed there needs to be a determination of how

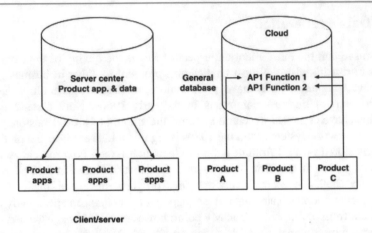

Fig. 8.3 Comparison of client/server and cloud architectures

applications and datasets are deployed. Obviously, there would be great advantage
to have the cloud resident on a quantum computer to support ML and AI processing
and improved cryptology.

There is no question that cloud is a sophisticated service-oriented architecture.
While many analysts and designers understand the concept of cloud, many do not
know how to maximize the configuration. Specifically, cloud should not be
designed as a client/server hierarchical and closely coupled system. Cloud must be
distributed, especially to support the new requirements of IoT. Therefore, cloud
architecture must parallel IoT needs and provide independent applications in the
form of functional primitives which will perform services independent of any given
system. Figure 8.3 depicts the difference between a client/server design and a cloud
distributed model.

The process of transforming existing systems to a cloud environment will be
discussed in Chap. 10; however, a preview of this issue is provided here for better
context. The first step in the transformation from legacy to cloud migration is to
"decouple:" the data from the legacy application system. Applications that own
their own data will not work well in a private or public cloud system. Once this
separation is completed, analysts need to determine where to place applications and
data on the network cloud systems. It is easier to replicate applications in multiple
places across the network but more complicated with data. Have both distributed
may have significant performance advantages particularly when it is time to
determine how much data should reside on devices on the Edge. Obviously, distribution of datasets will be very fundamental for a blockchain architecture than for
a more traditional client/server layout. Another issue will data distribution often
relates to sensitivity and policy decisions. Many companies may be sensitive to
having their data reside on a public cloud for instance. In most cases performance is
the significant decision maker which is still affected by the number of read and

write functions to and from databases that programs will perform during processing. Although many developers can use caching systems to improve performance, at some hardware latency will influence design decisions. Of course, having a quantum computer may greatly assist the latency issues depending on the type of processing being performed on the server. Overall, the mission for the analyst is to minimize input/output requests of all application programs. Always remember that the slowest operation on a computer remains the communication interaction among hardware devices. This design methodology is often known as designing for performance. In fact, studies have shown that overloading application server input output functions can deplete performance by over 80%! To address the latency potential, analyst should configure monitoring tools that can be used to alter load balances during peak processing.

Certainly, another variable in performance decisions is the role of security protection and its role in cloud analysis and design. I have already established that the world is moving to mobility, and cloud is a key part of a successful wireless infrastructure. However, we know with more mobility there is higher cyber exposure. Therefore, cloud applications should make use of "identity and access management." Completing design with security in mind is critical to secure systems and is often very dependent on industry risk protocols such as healthcare's "Health Insurance Portability and Accounting Act" or HIPAA compliance.

8.8 Cloud Architectures

Part of a successful mobile infrastructure is designing the right cloud architecture which depends on the business needs, the technology service requirements, and the available technological capabilities like quantum. As you can imagine depending on these variables, there are different cloud models. According to some of the excerpts from Architecting Cloud Computing Solutions there are three models to consider: baseline, complex, and hybrid.

Baseline

Baseline cloud computing is considered a foundational start as a beginner's cloud architecture. Baseline is a tiered and layered architecture with most having three basic tiers: web server, application layer, and database layer. Every tier has some amount if data storage that can vary based on the design requirements. Most cloud designs have some aspects of three tiers a shown in Fig. 8.4.

Within the baseline architecture there are various configurations:

Single Server

This design is hosted by a single server which could be virtual or physical and contains the three layers described above. This architecture is not recommended due of its security risks because one layer can compromise another. Because this design

Fig. 8.4 Three tier baseline
cloud architecture

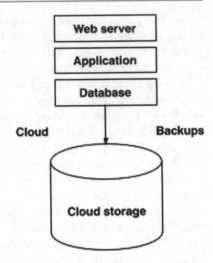

is inadequate for mobile deployments, it is usually limited to work as an internal development machine.

Single Site

This architecture has the same design as a single server except that each layer has its own computer instance and thus improves security, although all resources are still located on the same computer. There are two types of single-site architectures: non-redundant and redundant. Non-redundant architectures are essentially designed to save costs and resources but suffer from "single point of failure." Once again, while this option has multiple instances it is not recommended for production. Figure 8.5 reflects this design.

Redundant architecture on the other hand provides backup for failover and recovery protection. Thus, redundant design offers duplicate components that eliminate the single point of failure as shown in Fig. 8.6.

Obviously, redundant architecture are designed more for production systems because there are multiple processing decision capabilities that avoid single point of failure.

Complex Cloud Architectures

The complex cloud architecture addresses issues of redundancy, resiliency, and disaster recovery. At the core of complex cloud is the ability to monitor and adjust flow of traffic among multiple sites and to alternate balances appropriately based on usage. There are various types of complex cloud architectural designs.

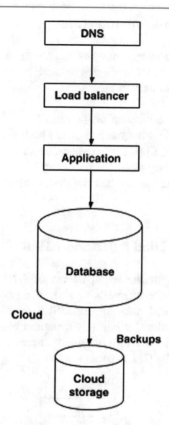

Fig. 8.5 Non-redundant three-tier architecture

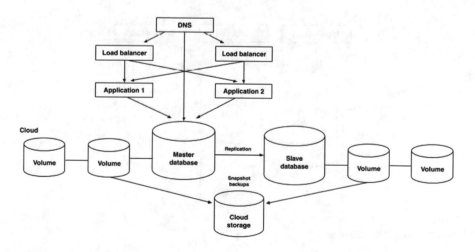

Fig. 8.6 Redundant three-tier architecture

8.8.1 Multi-data Center Architecture

A data center architecture allows analysts to determine the amount of redundant infrastructure needed to support single-site and multi-site designs. The major questions for the analyst to answer are:

- How is traffic sent to one location or the other?
- Is one site active and the other backup or are both active?
- How does fail-back to the primary site handled should a failure occur?
- What changes in resiliency plans are necessary?
- How is data synchronization handled before and after failover?

8.8.2 Global Server Load Balancing (GSLB)

This architecture allows for the manipulation of DNS (Domain Name Server) information. The DNS is the internet's version of a phonebook or address of the machine. Global server load balancing or GSLB enables pre-planned actions to occur in the event of a failure. While this design is effective, it is expensive and typically requires human interface. It is usually offered as a public cloud option for a fee. Figure 8.7 shows the GSLB configuration.

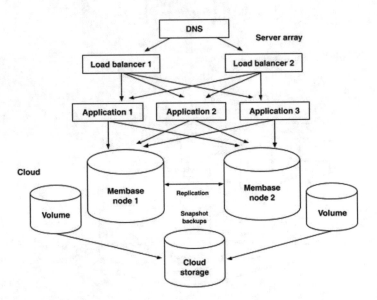

Fig. 8.7 GSLB architecture

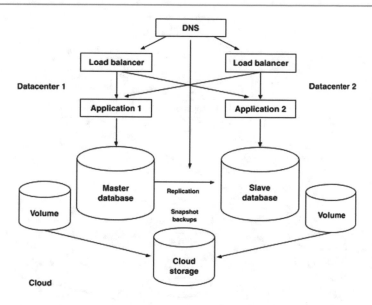

Fig. 8.8 Database resiliency architecture

8.8.3   Database Resiliency

This design offers what is called active-to-active database configuration with a bi-directional replication capability. This helps keep data synchronized on both database servers. While this design adds more complexity it also provides greater levels of redundancy and resiliency. Figure 8.8 shows the design.

Another option on databases its to add caching capabilities which holds data in high speed memory. The caching option works on algorithms that bet that certain data will be requested again. If that bet works it can significantly speed up data access. The idea behind caching is that an application may engage in multiple input and output operations for a period of time with the same records. Figure 8.9 shows the addition of caching memory.

8.8.4   Hybrid Cloud Architecture

Hybrid cloud is a solution that combines a private cloud with one or more public cloud services. Hybrid cloud provides greater flexibility because you can alter workloads among multiple cloud infrastructures. It also allows organizations to examine cost alternatives.

A hybrid cloud can certainly minimize exposure to a site failure because there are multiple failover options. It's clear in many ways that the hybrid option certainly is attractive for the IoT/Blockchain mobile operations because of the

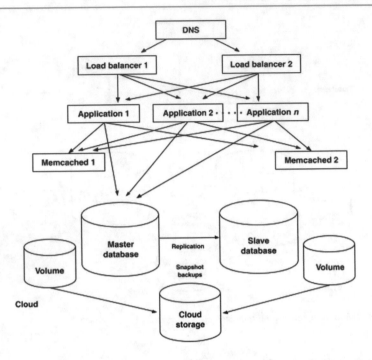

Fig. 8.9 Caching database cloud design

redundancy and multi-location load balances that it can offer. What is always true with sophisticated architectures is the higher costs, although using third party operators for competitive choices is part of the decision-making process. Beyond cost and failover is flexibility. Hybrid clouds allow owners to have that protection in a private cloud while offering the ability to extend onto a public cloud for more capacity as needed. The model is depicted in Fig. 8.10.

8.9 Cloud, Edge, and Fog

As IoT devices become widespread, organizations will need to store more data on devices, also known as the Edge. The edge devices and other network machines will need to interface with a more centralized cloud operation which have recently been coined Fog computing. The objective is to maximize performance and ensuring options for scalability especially during peak demands. Many organizations are considering collocating their IT infrastructure with other data centers to conserve costs. It is important to note that while Edge and the Cloud represent

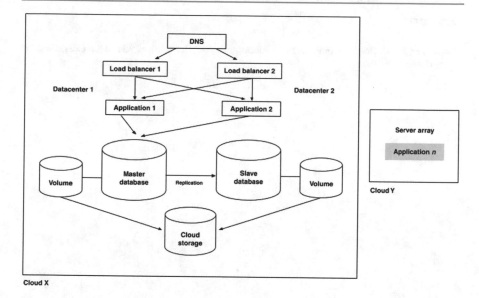

Fig. 8.10 Hybrid cloud architecture

current alternatives, the potential rise of quantum computing certainly offers an attractive addition to finding ways to store and analyze the incredible explosion of valuable consumer data.

8.10   Problems and Exercises

1. Define and describe quantum computing.
2. What are the advantages of quantum architecture?
3. How does quantum architecture relate to AI and ML? Be specific.
4. What is the relationship between quantum and hash keys?
5. What is a dataset? Describe the different types of sets.
6. Why is predictive analytics so dependent on AI and ML?
7. How do APIs increase performance of predictive analytics?
8. What are some of the disadvantages of ML?
9. What is Natural Language Processing and its relation to dataset?
10. Define two types of ML algorithms.
11. What are the challenges when updating data elements from legacy data?
12. What dilemma does ML create for database normalization?
13. What is Cloud? Why is it so essential for mobile-based architectures?
14. Compare client/server and cloud architectures.
15. Why is hybrid cloud architecture so attractive?
16. What is Fog?

Reference

Dorard, L. (2014). Bootstrapping machine learning: Exploit the value of your data. Create smarter apps and businesses.

Cyber Security in Analysis and Design

9

9.1 Introduction

The overall challenge in building more resilient applications that are better equipped to protect against threats is a decision that needs to address exposure coupled with risk. The general consensus is that no system can be 100% protected and that this requires important decisions when analysts are designing applications and systems. Indeed, security access is not just limited to getting into the system, rather on the individual application level as well. How then do analysts participate in the process of designing secure applications through good design? We know that many cyber security architectures are designed from the office of the chief information security officer or CISO, a new and emerging role in organizations. The CISO role, often independent of the CIO (chief information officer) became significant as a result of the early threats from the Internet, the 9/11 attacks and most recently the abundant number of system compromises experienced by companies such as JP Morgan Chase, SONY, Home Depot and Target to name just a few.

The challenge of cyber security reaches well beyond just architecture. It must address third-party vendor products that are part of the supply-chain of automation used by firms, not to mention access to legacy applications that likely do not have the necessary securities built into the architecture of these older, less resilient technologies. This challenge has established the need for an enterprise cyber security solution that addresses the need of the entire organization. This approach would then target third-party vendor design and compliance. Thus, cyber security architecture requires integration with a firm's Software Development Life Cycle (SDLC), particularly within steps that include strategic design, engineering, and operations. The objective is to use a framework that works with all of these components.

© Springer Nature Switzerland AG 2020
A. M. Langer, *Analysis and Design of Next-Generation Software Architectures*,
https://doi.org/10.1007/978-3-030-36899-9_9

9.2 Cyber Security Risk in the S-Curve

In Chap. 2, I discussed the importance of the S-Curve and its relationship to the risk in accuracy when determining requirements sources. The S-Curve also needs to be considered in the design of the cyber security architecture. When designing against cyber security attacks, as stated above, there is no 100% protection assurance. Thus, risks must be factored into the decision-making process. A number of security experts often ask business executives the question, "how much security do you want, and what are you willing to spend to achieve that security?"

Certainly, we see a much higher tolerance for increased cost given the recent significance of companies that have been compromised. This section provides guidance on how to determine appropriate security risks based on where the product exists on the S-Curve.

Security risk is typically discussed in the form of threats. Threats can be categorized as presented by Schoenfield (2015):

1. Threat Agent: where is the threat coming from and who is making the attack?
2. Threat Goals: what does the agent hope to gain.
3. Threat Capability: what threat methodology or type of is the agent possibly going to use?
4. Threat Work Factor: how much effort is the agent willing to put into get into the system?
5. Threat Risk Tolerance: what legal chances is the agent willing to take to achieve their goals?

Table 9.1 is shown as a guideline.

Depending on the threat and its associated risks and work factors will provide important input to the security design especially at the application design level. Such application securities in design typically include:

1. The user interface (sign in screen, access to specific parts of the application).
2. Command-line interface (interactivity) in on-line systems.
3. Inter-application communications. How data and password information are passed and stored among applications across systems.

Table 9.1 Threat analysis. *Source* Schoenfield (2015)

Threat agent	Goals	Risk tolerance	Work factor	Methods
Cyber criminals	Financial	Low	Low to medium	Known and proven

9.3   Decomposition in Cyber Security Analysis

Flow diagrams and database normalization is not the only application for functional decomposition. The concept of decomposition is very relevant to the architecture of systems that guard against attacks. Simply put, integrating security at the functional decomposed levels of applications can improve protection. However, security at this level does not negate threat exposure, but will reduce risk. Figure 9.1 shows a scheme of how security is implemented at all levels during the decomposition process.

Figure 9.1 demonstrates how analysts need to ensure that various security routines exist at multiple layers including the front-end interface, applications, and databases. The reason for this level of security is to provide the maximum amount of resistant to the ways threats enter a system. Indeed, once a virus enters it can find its way through many parts of the system—that is, entry points serve as gateways into the environment. Some of these points of potential entry could require special third-party applications, but these decisions should be discussed with the cyber

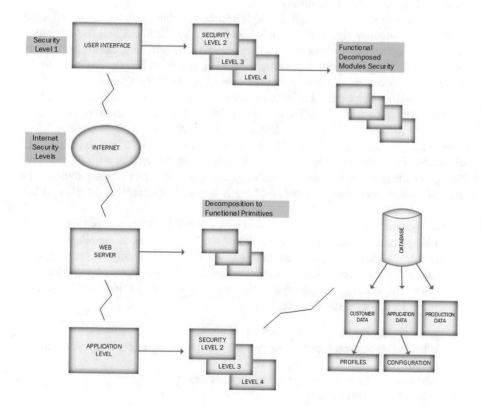

Fig. 9.1 Cyber security and decomposition

security team. The analyst must continue to stay focused on the logical view of the system.

At each decomposed level, the analyst should consider the specific checklist of threat potentials. The number of security interfaces at each level as well as the extent of the security needed will depend on the amount of risk and associated costs to implement. Important considerations prior to determine the levels of security is always the impact it might have on the user interface and on overall performance of the system.

9.4   Risk Responsibility

Schoenfield (2015) suggests that someone in the organization is assigned the role of the "risk owner." There may be many risk owners and as a result this role could have complex effects on the way systems are designed. For example, the top risk owner in most organizations today is associated with the CISO, or cyber security information systems officer. However, many firms also employ a CRO or chief risk officer. This role's responsibilities vary.

But risk analysis at the application design level requires different governance. Application security risk needs involvement from the business and the consumer and integrated within the risk standards of the firm. Specifically, multiple levels of security often require users to re-enter secure information. While this may maximize safety, it can negatively impact the user experience and the robustness of the system interface in general. Performance can obviously also be sacrificed given the multiple layers of validation. There is no quick answer to this dilemma other than the reality that more security checkpoints will reduce user and consumer satisfaction unless cyber security algorithms become more invisible and sophisticated. However, even this approach would likely reduce protection. As with all analysts, design challenges the IT team, business users, and now the consumer must all be part of the decisions on how much security is required especially in the 5G and IoT era.

As my colleague at Columbia University, Steven Bellovin states in his book, "Thinking Security," security is about a mindset. This mindset to me relates to how we establish security cultures that can enable the analyst to define organizational security as it relates to new and existing systems. If we get the analyst position to participate in setting security goals in our applications, some key questions according to Bellovin (2015) are:

1. What are the economics to protect systems?
2. What is the best protection you can get for the amount of money you want to spend?
3. Can you save more lives by spending that money?
4. What should you protect?
5. Can you estimate what it will take to protect your assets?

6. Should you protect the network or the host?
7. Is your Cloud secure enough?
8. Do you guess at the likelihood and cost of a penetration?
9. How do you evaluate your assets?
10. Are you thinking like the enemy?

The key to analysis and design in cyber security is recognizing that it is dynamic; the attackers are adaptive and somewhat unpredictable. This dynamism requires constant architectural change accompanied with increased complexity of how systems become compromised. Thus, analysts must be involved at the conceptual model, which includes business definitions, business processes and enterprise standards. However, the analysts must also be engaged with the logical design, which comprises two sub-models:

1. Logical Architecture: depicts the relationships of different data domains and functionalities required to manage each type of information in the system.
2. Component Model: reflects each of the sub-models and applications that provide various functions in the system. The component model may also include third-part vendor products that interface with the system. The component model coincides in many ways with the process of decomposition.

In summary, the analysis and design interface with cyber security is complex. It must utilize the high-level and decomposed diagrams and pictures that reference the specific hardware and software that comprise the needs for security. Security is relative, and thus analysts must be closely aligned with the CISO, executive management, and network architects to keep current with the threats and fixes when systems get compromised.

9.5 Developing a System of Procedures

Perhaps one of the most significant challenges in analysis today is its role in the software life cycle. There has been much criticism of the lack of discipline applied to software development projects and personnel in general, and we continue to be an industry that has a poor reputation for delivering quality products on schedule. This is even more evident with the exposures to cyber security attacks as articulated above. Although many organizations have procedures, few really follow them and fewer still have any means of measuring the quality and productivity of software development. A system of procedures should first be developed prior to implementing a life cycle that can ensure its adherence to the procedure. These procedures also need to be measured on an ongoing basis. This book restricts its focus to the set of procedures that should be employed in the analysis and design functions.

The process of developing measurable procedures in an organization must start with the people who will be part of its implementation. Standard procedures should not be created by upper management, as the steps will be viewed as a control mechanism as opposed to a quality implementation. How then do we get the implementers to create the standards? When examining this question, one must look at other professions and see how they implement their standards. The first main difference between computer professionals and members of many other professions is they lack a governing standards board like the American Medical Association (AMA) or the American Institute of Certified Public Accountants (AICPA). Unfortunately, as mentioned in previous chapters, it seems unlikely that any such governing board will exist in the near future. Looking more closely at this issue, however, we need to examine the ultimate value of a governing board. What standards boards really accomplish is to build the moral and professional responsibilities of their trade. Accountants, attorneys and doctors look upon themselves as professionals who have such responsibilities. This is not to imply that governing boards can resolve every problem, but at least they can help. With or without the existence of a standards board, analysts within an organization must develop the belief that they belong to a profession. Once this identification occurs, analysts can create the procedures necessary to ensure the quality of their own profession. There simply are not many analysts who view themselves as part of a profession.

If analysts can create this level of self-actualization, then the group can begin the process of developing quality procedures that can be measured for future improvement. The standard procedures should be governed by the group itself and the processes integrated into the software life cycle of the organization. In fact, analysts should encourage other departments to follow the same procedures for implementing their respective quality procedures.

9.6 IoT and Security

IoT and security are still in search of a good relationship. Specific challenges surround the security of pairing devices, encryption of links, registration and authentication of the device, updating of secrets, keys, and sensitive information in general. Thus, IoT with all of its advantages creates a unique set of challenges for security. While cyber security issues may vary in different geographies and industries, IoT's most significant security exposure relates to its decentralized architecture. Furthermore, since IoT is device-based, many of the cyber issues are in the hands of third-party vendors who build and support these devices. The result is that IoT is forcing companies to develop new security strategies. This security challenge is yet another activity for the analyst.

Unfortunately, there are no standards for securing IoT. This is especially disturbing given that public and private clouds operate over different sectors and countries. Certainly, another area of concern is how to tie in IoT security with legacy systems. A number of organizations have considered "retrofitting" which integrates old systems with new mobile-based architecture. Indeed, retrofitting is attractive given the alternative costs to build new systems. But keeping the old systems ultimately increases security exposure so companies need to assess the risks and explore all options.

However, one benefit of IoT decentralization is that it creates more loosely coupled systems that because of their independence reduce the likelihood of complete system failures. So, organizations within an IoT environment may only be faced with partial failures. Further, because of the redundancy that is built into IoT and blockchain architecture, analysts may actually have more alternative failover capabilities. This is also evident in Chap. 8 where I provided the number of cloud architectures, many of which are built to support partial failovers that are not just focused on system failures from power or hardware failures, but also against cyber security attacks.

But cyber is not limited to a case of shutdown, it's also about protecting data, particularly the data of consumers. In addition, various government lead laws like GDPR (General Data Protection Regulation)[1] governed by the European Union (EU), have stiff penalties for systems that are breached.

9.6.1 Cyber Security and Analyst Roles and Responsibilities

Operationally analysts need to consider taking on the following roles and areas of responsibility:

1. Apply service-oriented security architecture principles to meet the organization's confidentiality and integrity.
2. Ensure all security procedures are documented and updated properly and regularly.
3. Confirm that software patches and fixes are accomplished for both internal and external systems.
4. Ensure that all cyber products have identified risk acceptance levels.
5. Implement security countermeasures.
6. Perform testing on developed applications.
7. Conduct security reviews and identify gaps.
8. Make recommendations for better cyber design of systems.
9. Advise on disaster recovery, contingency, and continuity of operations.
10. Ensure every system has minimum security for all of its applications.
11. Participate in cyber recommendations based on the possibility of threats and vulnerabilities.

---

[1]GDPR is a regulation in EU law on data protection and privacy for all individual citizens of the EU and European Economic Area (EEA). It also addresses the transfer of personal data outside the EU and EEA areas (Wikipedia 2019).

In order to accomplish these roles and responsibilities, analysts will need the following skills:

- Knowledge of networking, protocols, and security methodologies.
- Knowledge of risk management both at the enterprise level as well as understanding different approaches to assessing and mitigating different risks.
- Knowledge of cyber laws, regulations, policies, and ethics (GDPR).
- Capabilities to apply Cyber security and privacy concepts.
- Knowledge of current cyber threats and reported vulnerabilities.
- Understanding of potential operational effects from cyber problems.
- Understanding of algorithms employed in cyber systems including IoT, blockchain, and cloud systems.
- Expertise in systems security testing and acceptance test planning.
- Knowledge of counter measurement for identified security risks.
- Knowledge of embedded systems.
- Awareness of network design processes including IoT interfaces, blockchain architecture and third-party products.
- Ability to use network tools to identify vulnerabilities.
- Ability to apply Cyber security and privacy principles to organizational requirements that relate to such areas as confidentiality, integrity, availability, authentication, and non-repudiation.
- Establish categories of the possible types of cyber-attacks and participate in the development of company-wide policies on cyber risk.

It should be obvious from the above tasks and responsibilities that the analysts have a lot of work to do! It should also be evident that this is not one person with capabilities in all the cyber areas. Thus, what I am really emphasizing is the need for a new organizational structure that has a team of analysts with specialized responsibilities similar to any other professional department. I will provide a more complete organization diagram of responsibilities in Chap. 13 (see Fig. 13.2).

9.7   ISO 9000 as a Reference for Cyber Standards

Although not typically required, many firms have elected to employ ISO 9000 as a more formal vehicle to implement the development of measurable procedures. ISO 9000 stands for the International Organization for Standardization, an organization formed in 1947 and based in Geneva, which currently has 91 member countries associated with it. ISO 9000 was founded to establish international quality assurance standards focused on processes rather than on products.

9.7.1   Why ISO 9000?

ISO 9000 offers a method of establishing agreed-upon quality levels through standard procedures in the production of goods and services. Many international companies require that their vendors be ISO 9000 compliant through the certification process. Certification requires an audit by an independent firm that specializes in ISO 9000 compliance. The certification is good for three years. Apart from the issue of certification, the benefits of ISO 9000 are in its basis for building a quality program through employee empowerment. It also achieves and sustains specific quality levels and provides consistency in its application. ISO 9000 has a number of sub-components. ISO 9001, 9002, 9003 codify the software development process. In particular 9001 affects the role of the analyst by requiring standards for design specifications and defines 20 different categories of systems. Essentially ISO 9000 requires three basic things:

1. Say What You Do.
2. Do What You Say.
3. Prove it.

This means that the analyst will need to completely document what should occur during the requirements process to ensure quality. After these procedures are documented the analyst needs to start implementing based on the standards developed and agreed upon by the organization. The process must be self documenting, that is, it must contain various control points that can prove at any time that a quality step in the process was not only completed but done within the quality standard established by the organization. It is important to recognize that ISO 9000 does not establish what the standard should be but rather simply that the organization can comply with whatever standards it chooses to employ. This freedom is what makes ISO 9000 so attractive. Even if the organization does not opt to go through with the audit, it can still establish an honorable quality infrastructure that:

- creates an environment of professional involvement, commitment and accountability.
- allows for professional freedom to document the realities of the process itself within reasonable quality measurements.
- pushes down the responsibilities of quality to the implementer as opposed to the executive.
- identifies where the analyst fits in the scope of the software life cycle.
- locates existing procedural flaws.
- eliminates duplication of efforts.
- closes the gap between required procedures and actual practices.
- complements the other quality programs that might exist.
- requires that the individuals participating in the process be qualified within their defined job descriptions.

9.8 How to Incorporate ISO 9000 into Existing Security and Software Life Cycles

A specific component of ISO 9000 is 9001 which focuses on the considerations of risk related to data availability and cyber security. There are also related standards such as ISO 27001 which addresses risk identification and mitigation processes. This includes the legal, physical, and technical controls needed in a risk information process. The question now is how to incorporate an ISO 9001-type process for the analyst function and incorporate it into the existing cyber security life cycle. Listed below are the essential 9 steps to follow:

1. Create and document all the quality procedures for the analyst.
2. Follow these processes throughout the organization and see how they enter and leave the analyst function.
3. Maintain records that support the procedures.
4. Ensure that all professionals understand and endorse the quality policy.
5. Verify that there are no missing processes.
6. Changes or modifications to the procedures must be systematically reviewed and controlled.
7. Have control over all documentation within the process.
8. Ensure that analysts are trained and that records are kept about their training.
9. Ensure that constant review is carried out by the organization or through third party audits.

Another component of ISO 9000 is ISO 27032 that provides guidelines cyber manager on:

- Data and privacy from threats
- Maintaining a cyber program
- Developing best practices
- Improving security systems and business continuity
- Building confidence of stakeholders
- Response and recovery from incidents.

In order for ISO 9000, 9001 and other cyber related guidelines to be implemented, it is recommended that the analyst initially provide a work flow diagram of the quality process as shown in Fig. 9.2.

Figure 9.2 reflects some of the steps in a quality process for an analyst. Note that certain steps reflect that there is an actual form that needs to be completed in order to confirm the step's completion as shown in Figs. 9.3, 9.4 and 9.5.

The above forms represent the confirmation of the activities in the quality work flow process outlined by the analyst. At any time during the life cycle, an event can be confirmed by looking at the completed form.

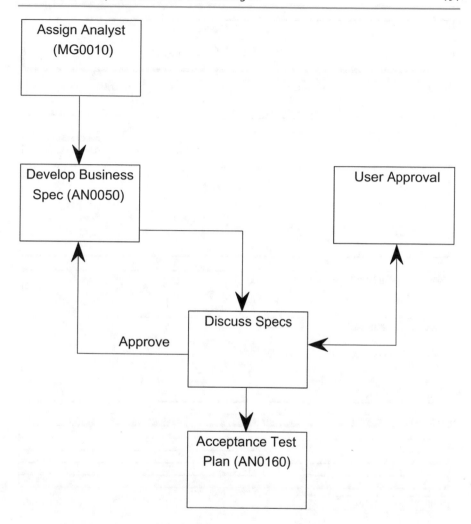

Fig. 9.2 Sample work flow diagram

In order to comply with the documentation standards, each form should contain an instruction sheet (as shown below). This sheet will ensure that users have the appropriate instructions. Confirmation documents can be implemented in different ways. Obviously if forms are processed manually, the documentation will contain the actual storage of working papers by project. Such working papers are typically filed in a documentation storage room similar to a library where the original contents are secure and controlled. Access to the documentation is allowed, but must be authorized and recorded. Sometimes forms are put together using a word-processing package such as Microsoft Word. The blank forms are stored on a central library so that master documents can be accessed by the analyst via a

Project Status Report

Period Ending: / / :

Date:	Project:	Analyst:
	Date Delivered To Users:	

Previous Objectives:

Objective	Previous Target Date	Act Completion Date or Status

New Project Objectives:

Objective	Target Date
Start Complete	

Financial Performance:

	Budget	Actual	% Remaining

\AN0010 Rev. 3/21/19

Fig. 9.3 ISO 9001 project status report

network. Once the forms are completed, they can be stored in a project directory. The most sophisticated method of implementing ISO 9000 is to use a Lotus Notes type electronic filing system. Here, forms are filled out and passed to the appropriate individuals automatically. The confirmation documents then become an inherent part of the original work flow. In any event, these types of forms implementation affect only automation, not the concept of ISO 9000 as a whole as shown in Fig. 9.6.

Analysis Acknowledgment

User: _____

Date: _____   Analyst: _____   Date of Request _____   Project # _____

Confirmation Type	Y/N	Cost	Expected Days	Expected Delivery	Deliverable / Comments
Requirements Definition					
Conceptual Detail Design:					
Development:					
System Tested Enhancements					
User Accepted Enhancements					

\an0050 Rev. 3/21/19

Fig. 9.4 ISO 9000 analysis acknowledgment

**Quality Assurance**

**Acceptance Test Plan**

Purpose:

Test Plan #:

Product:

Number:

Vendor:

Date:

QA Technician:

Page: 1 of

Test No.	Condition Being Tested	Expected Results	Actual Results	Comply Y/N	Comments
1					
2					
3					

Fig. 9.5 Quality assurance test plan

Name:	Date Issued: 3/8/19
Confirmation/Service	
Acknowledgment	
	Supersedes:
Form Instructions	
	Revision: 1.00

The purpose of this form is to track the status of various services such as Requirements Definition, Conceptual Detail Design, Development, Cyber System Tested Enhancements and User Accepted Enhancements.

The appropriate project # must be attached. The type form must be checked for each type of Confirmation.

\an0050i Rev 3/8/11

Fig. 9.6 An ISO 9000 form instruction page

9.9   Interfacing IT Personnel

We mentioned earlier that ISO 9000 requires qualified personnel. This means that the organization must provide detailed information about the skill set requirements for each job function. Most organizations typically have job descriptions that are not very detailed and tend to be vague with respect to the specific requirements of the job. In addition, job descriptions rarely provide information that can be used to measure true performance. Questions such as "How many lines of code should a programmer generate per day?" cannot be measured effectively. There is also a question about whether lines of code should be the basis of measurement at all. A solution to this dilemma is to create a Job Description Matrix which provides the specific details of each job responsibility along with the necessary measurement criteria for performance as depicted in Fig. 9.7.

The document above is a matrix of responsibilities for an analyst. Note that the analyst has a number of efficiency requirements within the managing engagements (projects) responsibility. Efficiency here means that the analyst must perform the task at a certain indicated level to be considered productive at that task. To a degree, efficiency typically establishes the time constraints to deliver the task. Measurement defines the method used to determine whether the efficiency was met. Reports are

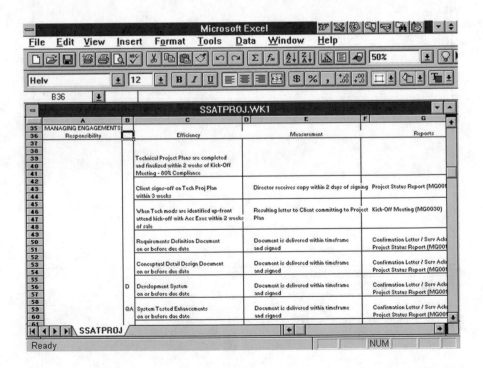

Fig. 9.7 Job description matrix

simply the vehicle in which the analyst proves that the task was completed and on what basis.

The Job Description Matrix represents a subset of the entire job description that strictly focuses on the procedural and process aspects of the individual's position. It not only satisfies ISO 9000, but represents a healthier way of measuring individual performance in an IS environment. Most individuals should know exactly their performance at any time during the review period. Furthermore, the matrix is an easy tool to use for updating new or changed performance tasks.

9.10   Committing to ISO 9000

We have outlined the sequential steps to implement an ISO 9000 organization. Unfortunately, the outline does not ensure success, and often just following the suggested steps leads to another software life cycle that nobody really adheres to. In order to be successful, a more strategic commitment must be made. Let's outline these guidelines for the analyst functions:

- A team of analysts should meet to form the governing body that will establish the procedures to follow to reach an ISO 9000 level (this does not necessarily require that certification be accomplished.)
- The ISO 9000 team should develop a budget of the milestones to be reached and the time commitments that are required. It is advisable that the budget be forecasted like a project, probably using a Gantt chart to develop the milestones and timeframes.
- The ISO 9000 team should then communicate their objectives to the remaining analysts in the organization and coordinate a review session so that the entire organization can understand the benefits, constraints and scope of the activity. It is also an opportunity to allow everyone to voice their opinions about how to complete the project. Therefore, the meeting should result in the final schedule for completing the ISO 9000 objective.
- The ISO 9000 team should inform the other IS groups of its objectives, although analysts should be careful not to provoke a political confrontation with other parts of the IS staff. The communication should be limited to helping other departments understand how these analyst quality standards will interface with the entire software life cycle.
- The work flows for the analyst tasks must be completed in accordance with the schedule such that everyone can agree to the confirmation steps necessary to validate each task. It is important that the ISO 9000 processes allow for a percentage of success. This means that not every process must be successful 100% of the time, but rather can be acceptable within some fault tolerance level. For example, suppose that the analyst must have a follow-up meeting with the users within 48 h after a previous step has been completed. It may not be realistic to

meet this goal every time such a meeting is necessary. After all, the analyst cannot always force users to attend meetings in a timely way. Therefore, the ISO 9000 step may view this task as successful if it occurs within the 48 h 80% of the time, that is, within a 20% fault tolerance.

- All task steps must have verification. This will require that standard forms be developed to confirm completion. While we have shown samples of these forms earlier, the ISO 9000 team should beware of producing an unwieldy process involving too many forms. Many software life cycles have suffered the consequences of establishing too many check points. Remember, ISO 9000 is a professional's standard and should cater to the needs of well-trained professionals. Therefore, the ISO 9000 team should review the initial confirmation forms and begin the process of combining them into a smaller subset. That is, the final forms should be designed to be as generic as possible by confirming multiple tasks. For example, form AN0010 (Fig. 9.3) represents a generic project status report that was used to confirm various types of information attributable to different tasks.
- There should be meetings held with the analysis group that focus on the alternatives for automating the confirmation forms as outlined earlier in this chapter. It is advisable that this topic be confirmed by the group since their full cooperation is needed for the success of the program.
- Allow time for changing the procedures and the forms. Your first effort will not be the final one; therefore, the ISO 9000 team must plan to meet and review the changes necessary to make it work. Analysts should be aware that the opportunity for change always exists as long as it conforms to the essential objectives of ISO 9000.
- The ISO 9000 project should be at least a one-year plan, from inception of the schedule to actual fulfillment of the processes. In fact, an organization must demonstrate ISO 9000 for at least 18 months prior to being eligible for certification.
- The ISO 9000 group needs to be prepared and authorized to make changes to the job description of the analyst. This may require the submission of requests and authorizations to the executive management team or the human resources department. It is important not to overlook this step since an inability to change the organization structure could hinder the success of the ISO 9000 implementation.

As we can see from the above steps, establishing an ISO 9000 group is a significant commitment. However, its benefits can include a professional organization that controls its own quality standards. These standards can be changed on an ongoing basis to ensure compliance with the business objectives and requirements of the enterprise. Certification, while not the focus of our discussion, is clearly another level to achieve. Most companies that pursue certification do so for marketing advantage or are required to obtain it by their customers. Implementing ISO 9000 should not require that the entire company conform at once; in fact, it is almost an advantage to implement it in a phased approach, department by department. The potential benefits of ISO 9000 concepts may fill the void in many of the IS organizations which lack clearly defined quality standards.

9.11   Problems and Exercises

1. Why is the CISO role so important in mobile systems.
2. Why is the s-curve important when determining Cyber security risk?
3. What is threat analysis?
4. Explain the relationship between Cyber security and decomposition.
5. What is risk responsibility?
6. What is GDPR and its importance in data protection?
7. Explain the relationship between IoT and Cyber security.
8. List and define five roles and responsibilities of the Cyber security risk analyst.
9. Provide the key issues when designing systems to deal with cyber security attacks.
10. How does the analyst role change when being engaged with cyber security design?
11. Explain why ISO 9000 represents a system of procedures.
12. What are the three fundamental things that ISO 9000 tries to establish?
13. What are the overall benefits of ISO 9000?
14. How is ISO 9000 incorporated into the life cycle?
15. What is ISO 27001 and 27032?
16. Why are work flows the most critical aspect of developing the ISO 9000 model?
17. Why are forms used in ISO 9000?
18. How are personnel affected by ISO 9000?
19. What is a Job Description Matrix?
20. What steps are necessary for an organization to adopt ISO 9000?
21. Does ISO 9000 need to be implemented in all areas of the business? Explain.

References

Bellovin, S. M. (2015). To appear. *Thinking security: Stopping next year's hackers*. Boston: Addison-Wesley.
Schoenfield, B. S. E. (2015). *Securing systems: Applied security architecture and threat models*. Boca Raton, FL: CRC Press.
Wikipedia (2019).

Transforming Legacy Systems

10

10.1 Introduction

A Legacy system is an existing application system in operation. While this is an accurate definition, there is a perception that legacy systems are old and antiquated applications operating on mainframe computers. Indeed, Brodie and Stonebraker (1995) state that, "a legacy information system is any information system that significantly resists modification and evolution" (p. 3). They define typical legacy systems as:

- Large application with millions of lines of code.
- Usually more than 10 years old.
- Written in legacy languages like COBOL.
- They are built around a legacy database service (such as IBM's IMS) and some do not use a database management system. Instead they use older flat-file systems such as ISAM and VSAM.
- The applications are very autonomous. Legacy applications tend to operate independently from other applications, which means that there is very limited interface among application programs. When interfaces among applications are available, they are usually based on export and import models of data and these interfaces lack data consistency.

While many legacy systems do fit these scenarios, many do not. Those that do not can; however, be considered legacies under my original definition, that is, any application in operation. What this simply means is that there are what I call "generations" of legacy systems that can exist in any organization. Thus, the definitions of what constitutes a legacy system are much broader than Brodie and Stonebraker's descriptions. The more important issue to address is the relationship of legacy systems with packaged software systems especially with the explosion of independent APIs from the evolving IoT devices, blockchain products and Cloud computing. Packaged software systems that are typically supported by third-party

© Springer Nature Switzerland AG 2020
A. M. Langer, *Analysis and Design of Next-Generation Software Architectures*,
https://doi.org/10.1007/978-3-030-36899-9_10

vendors encompass both internal and external applications. Thus, existing internal production systems, including third-party outsourced products, must be part of any application strategy. Furthermore, there are many legacy systems that perform external functions as well, albeit not in directly in an internet interface.

This chapter defines the type of legacy systems that exist and provides guidelines on how to approach their integration with packaged software applications and transformation in into the new architectures that support IoT. The project manager or analyst must determine whether a legacy system should be replaced, enhanced, or remain as is. This chapter also provides the procedures for dealing with each of these three choices and its effect on the overall architecture of the decision whether to make or buy a system. However, overall this chapter suggests that all traditional legacy at some point need to redeveloped to support mobility and maximized cyber protection.

10.2   Types of Legacy Systems

The types of legacy systems tend to mirror the life cycle of software development. Software development is usually defined within a framework called "generations." Most professionals agree that there are five generations of programming languages:

1. *First-Generation*: the first generation was known as machine language. Machine language is considered a low-level language because there it uses binary symbols to communicate instructions to the hardware. These binary symbols form a one-to-one relationship between a machine language command and a machine activity, that is, one machine language command performs one machine instruction. It is rare that any legacy systems have first-generation software.
2. *Second-Generation*: this generation was comprised of assembler programming languages. Assembler languages are proprietary software that translates a higher-level coding scheme into more than one machine language instruction. Therefore, it was necessary to design an assembler, which would translate the symbolic codes into machine instructions. Mainframe shops still may have a considerable amount of assembly code that exists, particularly with applications that perform intricate algorithms and calculations.
3. *Third-Generation*: these languages continued the growth of high-level symbolic languages that had translators into machine code. Examples of third-generation languages are COBOL, FORTRAN, BASIC, RPG, and C. These languages use more English-like commands and have a higher ratio of machine language produced from one instruction. Third-generation languages tend to be more specialized. For example, FORTRAN is better suited for mathematical and scientific calculations. Therefore, many insurance companies have FORTRAN because of the high concentration on actuarial mathematic calculations. COBOL, on the other hand, was designed as the business language and has special features to allow it to manipulate file and database information. There are

more COBOL applications that exist than any other programming language. Most mainframe legacy systems still have COBOL applications. RPG is yet another specialized language that were designed for use on IBM's mid-range machines. These machines include the System 36, System 38, and AS/400 computers.

4. *Fourth-Generation (4GL)*: these programming languages are less procedural than third-generation languages. In addition, the symbols are more English-like and emphasize more about desired output results than how the programming statements need to be written. As a result of this feature, many less technical programmers have learned how to program using 4GLs. The most powerful features of 4GLs include query of databases, code-generation, and graphic screen generation abilities. Such languages include Visual Basic, C ++, Visual Basic, Powerbuilder, Delphi, and many others. Furthermore, 4GL languages also include what is known as *query languages* because they contain English-like questions that are used to directly produce results by directly accessing relational databases. The most popular 4GL query language is Structured Query Language (SQL).

5. *Fifth-Generation*: these programming languages combine what is known as rule-based code generation, component management, and visual programming techniques. The rule-based code generation became popular in the late 1980s with the creation of artificial intelligence software. This software uses an approach called knowledge-based programming, which means that the developer does not tell the computer how to solve problems, but rather the problem (Stair and Reynolds 1999). The program figures out how to solve the problem. While knowledge-based programming has become popular in specialized applications, such as in the medical industry, it has not been as popular in business.

Most legacy applications will either be third- or fourth-generation language systems, therefore, analysts need to have a process and methodology to determine how to transform and re-architect these applications.

10.3 Third-Generation Language Legacy System Integration

As previously discussed, most third-generation language legacy systems were developed using COBOL. COBOL was developed to provide a method of forcing programmers to self-document their code so that other programmers could maintain it. Unfortunately, COBOL requires what is known as a File Description Table (FD). The FD defines the record layout for every file used by the COBOL program. In other words, every file is described within the program and must match the format of the actual physical data file. This means that any change to a file structure must be synchronized with every COBOL program that uses that data file. Thus, COBOL is somewhat eclectic: there is no real separation of the data description and the program logic. In COBOL programs then, a change in data format could necessitate a change in the program code. That is why COBOL programs suffer from large degrees of coupling of code. Coupling is defined as the reliance of one piece of the code on another.

COBOL programs may or may not use a relational database as its source of data. I earlier defined two other common formats called ISAM and VSAM, which are flat-file formats, meaning that all data elements are contained in one record as opposed to multiple files as is the case in the relational database model. However, many COBOL legacy systems have been converted to work with relational databases such as IBM's DB2. In this situation the FD tables interface with the database's file manager so that the two entities can communicate with each other. Figure 10.1 depicts the interface between program and database.

Notwithstanding whether the COBOL legacy is using a database interface or flat-files the analyst needs to determine whether to replace the application, enhance it, or leave it as is.

10.4 Replacing Third-Generation Legacy Systems

When replacing third-generation legacies, analysts must focus on both the data and processes. Because of the age of these systems, it is likely that there will be little documentation available, and the amount available will most likely be outdated. Indeed, the lack of proper documentation is the major reason why legacy systems are slow to be replaced: rewriting code without documentation can be an overwhelming and time-consuming task. Unfortunately, all things must eventually be replaced. Delaying replacement leads to legacy systems that keep businesses from remaining competitive. The following sections provide a step-by-step approach to COBOL-based legacies.

10.5 Approaches to Logic Reconstruction

The best way to reconstruct the logic in COBOL applications is to separate the data from the processes. This can be accomplished by creating data flow diagrams (DFD) for each program. Having a DFD will result in defining all of the inputs and outputs of the application. This is accomplished by following the tasks below:

1. Print the source code (actual COBOL written code) from each application. Each application will contain a "FD" section that defines all of the inputs and outputs of the program. These will represent the data stores of the data flow diagrams (Fig. 10.2).

Employee Payroll File Description Table in COBOL Program

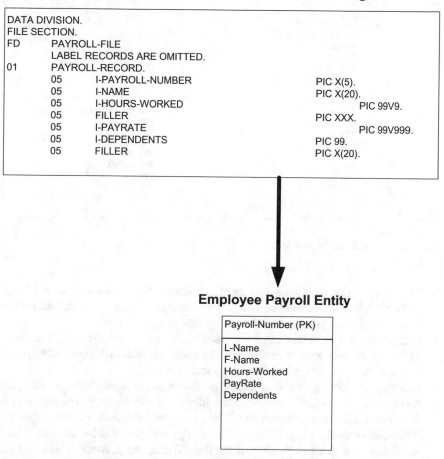

```
DATA DIVISION.
FILE SECTION.
FD      PAYROLL-FILE
        LABEL RECORDS ARE OMITTED.
01      PAYROLL-RECORD.
        05      I-PAYROLL-NUMBER            PIC X(5).
        05      I-NAME                      PIC X(20).
        05      I-HOURS-WORKED                      PIC 99V9.
        05      FILLER                      PIC XXX.
        05      I-PAYRATE                           PIC 99V999.
        05      I-DEPENDENTS                PIC 99.
        05      FILLER                      PIC X(20).
```

Employee Payroll Entity

Payroll-Number (PK)
L-Name F-Name Hours-Worked PayRate Dependents

Fig. 10.1 COBOL file description interface with database manager

2. DFDs should be decomposed so that they are at the functional primitive level (one-in and one-out, preferred). This provides functional decomposition for the old application and sets the framing for how it will be decomposed into an object-oriented solution.
3. By reviewing the code, write the process specifications for each functional primitive.
4. Follow the steps as outlined in Chap. 3 to determine which functional primitive DFDs become methods of a particular class.
5. Capture all of the data elements or attributes required by each functional primitive DFD. These attributes are added to the data dictionary (DD).

```
FD      REPORT-FILE
        LABEL RECORDS ARE OMITTED.
01      REPORT-RECORD.
        05      O-PAYROLL-NUMBER                PIC X(5).
        05      FILLER                          PIC XX.
        05      O-NAME                          PIC X(20).
        05      FILLER                          PIC XX.
        05      O-HOURS-WORKED                  PIC 99.9.
        05      FILLER                          PIC XX.
        05      O-PAYRATE                       PIC 99.999.
        05      FILLER                          PIC XX.
        05      O-DEPENDENTS                    PIC 99.
        05      FILLER                          PIC XX.
        05      O-GROSS-PAY                     PIC 999.99.
        05      FILLER                          PIC XX.
        05      O-TAX                           PIC 999.99.
        05      FILLER                          PIC XX.
        05      O-NET-PAY                       PIC 999.99.
```

Fig. 10.2 COBOL file description tables

6. Take each major data store and create an entity. Do normalization and Logic Data Modeling in accordance with the procedures in Chap. 4, combining these elements with the packaged software system as appropriate.
7. Data stores that represent reports should be compared against sample outputs. These reports will need to be redeveloped using a report-writer such as Crystal's report writer or a data warehouse product.
8. Examine all existing data files and/or databases in the legacy system. Compare these elements against those discovered during the logic reconstruction. In third-generation products there will be many data elements or fields that are redundant or used as logic "flags." Logic flags consists of fields used to store a value that depicts a certain state of the data. For example, suppose a record has been updated by a particular program. One method of knowing that this occurred is to have the application program set a certain field with a code that identifies that it has been updated. This method would not be necessary in a relational database product because file managers automatically keep logs on the last time a record has been updated. This example illustrates how different third-generation legacy technology differs from more contemporary technologies.

There is no question that replacing third-generation legacies is time-consuming. However, the procedures outlined above will prove to be accurate and effective. In many situations, users will decide that it makes sense to re-examine their legacy processes, especially when the decision has been made to rewrite applications for integration with IoT and blockchain systems. We call this business process reengineering. Business process reengineering is therefore synonymous with enhancing the legacy system.

10.6   Enhancing Third-Generation Legacy Systems

Business process reengineering (BPR) is one of the more popular methodologies used to enhance third-generation applications. A more formal definition of BPR is "a requirement to study fundamental business processes, independent of organization units and information systems support, to determine if the underlying business processes can be significantly streamlined and improved."[1] BPR is not just rebuilding the existing applications for the sake of applying new technology to older systems, but also an event that allows for the application of new procedures designed around the Object-Oriented systems paradigm. In this scenario, however, BPR is used to enhance existing applications without rewriting them in another generation language. Instead, the analyst needs to make changes to the system that will make it function more like an object component even though it is written in a third-generation language. In order to accomplish this task, it is necessary for the analyst to create the essential components of the legacy operation. Essential components represent the core business requirements of a unit. Another way of defining core business requirements is to view essential components as the reasons why the unit exists—what does it do—and for what reasons. For example, Fig. 10.3 depicts the essential components of a bank.

Once the essential components have been created then the legacy applications need to be placed in the appropriate component so that it can be linked with its related packaged software applications or decomposed into primitive APIs.

The first step to applying successful BPR to legacy applications is to develop an approach to defining the existing system and extracting its data elements and applications. Once again, this is similar to the process described above when replacing third-generation legacy applications in that the data needs to be captured into the data repository and the applications need to be defined and compared to a new model based on essential components.

10.7   Data Element Enhancements

The analyst will need to design conversion programs that will access the data files that are not in relational database format and place them in a data repository. The ultimate focus is to replace all of the existing data files in the legacy with relational databases that can be linked with packaged software databases. This methodology differs from replacing legacies. In replacement engineering, data files are integrated directly into a packaged software system. This means that the legacy data will often be used to enhance the packaged software databases or integrated with various cloud database products. However, the process of enhancing legacy systems means that the legacy data will remain separate but converted into the relational or object

---

[1]Whitten (2000).

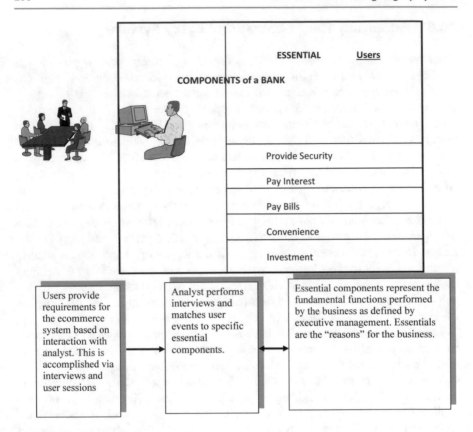

Fig. 10.3 Essential components of a bank

database model. For legacies that already have relational databases, there is no restructuring required beyond setting up links with the packaged software database. Figure 10.4 reflects the difference between replacing legacy data and enhancing it. Notwithstanding these steps, once these data elements are determined, the analyst should follow the steps to consider what elements should be considered for replication on IoT and blockchain architectures.

Another interesting difference between the two approaches is that enhanced legacies will likely have intentional data redundancy. This means that the same element may indeed exist in multiple databases; necessary for the new distributed systems supported by IoT and blockchain. Duplicate elements may take on different formats. The most obvious is where a data element has alias', meaning that an element has many different names, but the same attributes. Another type is the same element name, but with different attributes. The third type, and the most challenging, is the duplicate elements that have different names and different attributes. While duplicate data elements may exist in enhanced legacy applications that are integrated with the packaged software product, it is still important to identify

Fig. 10.4 Replacing versus enhancing legacy data

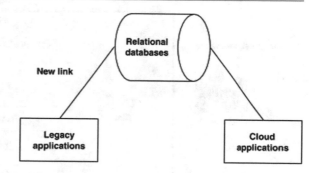

duplicate data relationships. This can be accomplished by documenting data relationships in a CASE tool and in the database's physical data dictionary where aliases can exist.

10.7.1   Application Enhancements

BPR typically involves a methodology called Business Area Analysis (BAA). The purpose of BAA is to:

- establish the various legacy business areas that will be linked with new architectures and/or packaged software systems.
- re-engineer the new and old requirements of each business area.
- develop requirements that provide an OO perspective of each legacy business area, meaning that there is no need to map its requirements to the existing physical organization structure.
- define the links that create relationships among all the legacy business areas and the packaged software business areas.

This is accomplished by mapping business areas to specific essential component. Applications designed for the packaged software system must also be mapped to an essential component. Once this has occurred, the legacy applications and packaged software applications must be designed to share common processes and databases as shown in Fig. 10.5.

Once the legacy and packaged software applications have been placed in their appropriate essential component they will need to be linked, that is communicate with each other to complete the integration of the internal IoT and external systems. Linking occurs in two ways: parameter messaging and database. Parameter messaging requires that the legacy programs be modified to receive data in the form of parameters. This allows the application system to deliver information directly to the legacy program. Conversely, the legacy program may need to return information back to the packaged software system. Therefore, legacy applications need to be enhanced so they can actually format and send a data message to the packaged

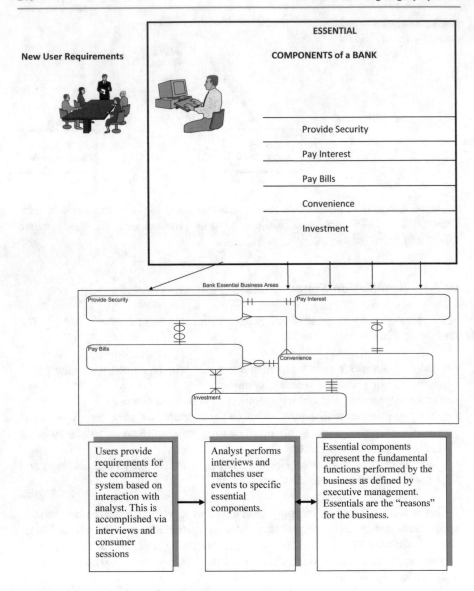

Fig. 10.5 BPR legacy modeling using essential components

software system. A database interface is essentially the same concept except that it occurs differently. Instead of the application sending the data directly to another program, it forwards it as a record in a database file. The legacy program that returns the data also needs to be modified to forward messages to a likely cloud database.

There are advantages and disadvantages of using either method. First, parameters use little overhead and are easy to program. They do not provide reusable data, that is, once a message has been received it is no longer available to another program. Parameters are also limited in size. Databases, on the other hand, allow programs to send the information to multiple destinations because it can be read many times. Unfortunately, it is difficult to control what applications or queries can access the data, which does raise questions about how secure the data is. Furthermore, applications must remember to delete a record in the database if it is no longer required. Figure 10.6 reflects the two methods of transferring data between legacy systems and packaged software applications.

Once the legacy and the new applications have all been mapped to the essential components, analysts can use a CRUD diagram to assist them in reconciling whether all of the data and processes have been found. The importance of the CRUD diagram is that it ensures:

- that an essential component has complete control over its data,
- that all of the entities are accessible by at least one process, and
- that processes are accessing data (Fig. 10.7).

While CRUD is not 100% accurate, it certainly uncovers potential problems as shown above. Even if BPR is not used, the CRUD diagram is an excellent tool to use to determine the processes and data needed for an essential component or an object. Once the CRUD diagram is finalized, the objects and classes would then be created as shown in Fig. 10.8; some of these objects are still in the form of a third-generation COBOL program while others might be in an API-based packaged software format. It is important to note; however, the "U" and the "D" are not allowed in a blockchain application, because ledger systems do not allow modification of existing transactions!

10.8   "Leaving As Is"—Third-Generation Legacy Systems

Moving to an object-oriented and API paradigm from a third-generation product like COBOL may not be feasible. The language design of third-generation procedural programs may result in conceptual gaps between procedural and object-oriented philosophies. For example, distributed object-oriented programs require more detailed technical infrastructure knowledge and graphics manipulation then was required in older legacy systems. Native object-oriented features such as inheritance, polymorphism, and encapsulation do not apply in traditional third-generation procedural design. It is difficult, if not impossible to introduce new object concepts and philosophies during a direct COBOL to JAVA API migration. If the translation is attempted without significant restructuring (as discussed earlier in this chapter) then the resulting product will likely contain slower code that is more difficult to maintain.

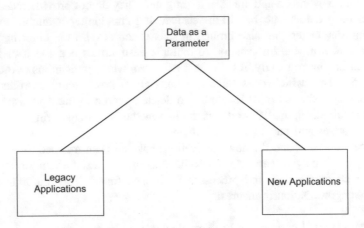

Legacy Link Using a Parameter

Legacy Link using a Database

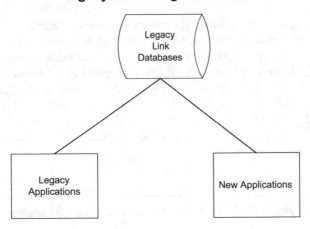

Fig. 10.6 Linking data between legacy and packaged software systems

There can also be a cultural divide that occurs. Veteran COBOL programmers and newer JAVA API developers do not understand each other's technologies. This scenario will often create bias during any conversion effort. Furthermore, COBOL programmers learning new technology can present them with self-specified threats to their cultural status. Furthermore, COBOL and RPG applications have benefited from more lengthy testing, debugging, and overall refinement than newer programming generations. While JAVA is more dynamic, it is less stable, and the procedure of debugging and fixing problems are very different than for COBOL or RPG. Therefore, the analyst will leave the legacy "as is," and create only packaged

Processes or Business Function / Data Subject or Entity	Process Orders	Validate Products	Shipping	Commission
Customers	R		R,U	
Orders	C, U		U	R
Items	R	C,U,D	R	R
Inventory	R,U	C,U	U	
Expense Section				
Market Person	R			U

Fig. 10.7 Sample CRUD diagram

software links for passing information that is required between the two systems. While this is similar to the "linking" proposed for enhancing legacy systems, it is different because legacy programs are not enhanced, except for the external links needed to pass information. This is graphically shown in Fig. 10.9.

Using parameters or databases to link connecting information is still relevant, but analysts must be cognizant that legacy data formats will not be changed. This means that the legacy applications will continue to use their original file formats. Another concept used to describe "links" is called "bridges." The word suggests that the link serves to connect a gap between the packaged software system and the legacy applications. Bridging can also imply temporary link. Very often "as is" can be seen as a temporary condition because legacy conversions cannot occur all at once, so it is typically planned in phases. However, while parts of the system are being converted, some portions may need temporary bridges until it is time to actually enhance them. One can visualize this as a temporary "road block" or detour that occurs when there is construction on a highway.

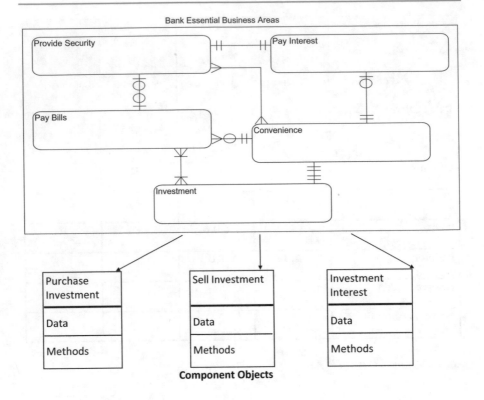

Fig. 10.8 Essential component object diagrams

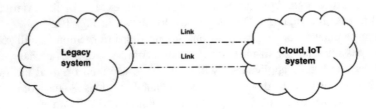

Fig. 10.9 "As Is" legacy links

10.9   Fourth-Generation Language Legacy System Integration

Integrating fourth-generation legacy systems with packaged software technology is much easier than third-generation languages. The reasons are two-fold. First, most fourth-generation implementations are already using a relational database, so conversion of data to the packaged software system is less complex. Second, fourth-generation language applications typically use SQL-based code, so conversion to an object-oriented system is also less involved.

10.10   Replacing Fourth-Generation Legacy Systems

As stated above, replacement of fourth-generation language systems is less complex than third-generation languages with respect to the packaged software conversion. As with any system replacement, separating data and process is the suggested approach. Fortunately, in fourth-generation language systems, process and data are likely to already be separate, because of the nature of its architecture. Specifically, fourth-generation languages typically use relational databases, which architecturally separate data and process. Therefore, replacing the legacy is more about examining the existing processes and determining where the applications need to be reengineered.

10.11   Approaches to Logic Reconstruction

The best approach to logic analysis is to print out the source code of the programs. If the source is written in SQL, then the analyst should search for all SELECT statements. SQL SELECT FROM statements define the databases that the program is using as shown in Fig. 10.10.

As in third-generation languages logic reconstruction, the analyst should produce a DFD for every program as follows:

1. SELECT statements define all inputs and outputs that a program uses. Each SELECT statement file will be represented by a DFD data store. Reviewing the logic of the application program will reveal whether the data is being created, read, updated, or deleted (CRUD).
2. DFD's should be decomposed to the functional primitive level so that the framework to an object-oriented system is established.
3. For each DFD copy the relevant SQL code, making modifications where necessary to provide more object-oriented functionality to the program. This

Select IdNo, Last-Name, First-Name
from   employee
where IDNo = "054475643"

Fig. 10.10 SELECT statements in a fourth-generation language application

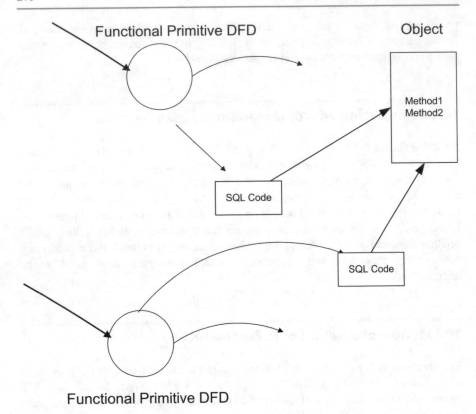

Fig. 10.11 SQL code transition to object method

means that the decomposition of the code will likely require that some new logic be added to transform it to a method. This is shown in Fig. 10.11.

4. Examine existing system objects and determine if functional primitive DFDs belong to an existing class as a new method, or whether it truly represents a new object in the packaged software system.

5. Capture all of the data elements required by the new methods and add them to their respective object. Ensure that the packaged software DD is updated appropriately.

6. Determine whether any new objects need to become a reusable component in the TP monitor (middleware), a reusable component in the client application, or as a stored procedure at the database level.

7. Examine the legacy databases and do logic data modeling to place the entities in third-normal form (3NF).

8. Combine and integrate data elements with packaged software databases, ensuring that each data field from the legacy system is properly matched with

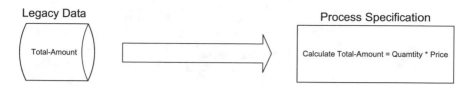

Fig. 10.12 Transition of redundant data elements to process specifications

packaged software data elements. New elements must be added to the appropriate entity or require that new entities be created for them.

9. Link new entities with existing models using third-normal form referential integrity rules.
10. Determine which data elements are redundant, such as calculations. These data elements will be removed; however, logic to their calculations may need to be added as a method as shown in Fig. 10.12.

10.12   Enhancing Fourth-Generation Legacy Systems

Enhancing fourth-generation language legacy systems is really the process of converting it to an object-oriented client/server system. BPR (Business Process Reengineering) is also used on fourth-generation language legacy systems to accomplish this transition. The process, as one might expect, is much easier than for third-generation languages; however, the process of determining essential components is the same in both type of systems. Once essential components are established, the existing applications need to be decomposed and realigned as appropriate. This is accomplished by using BAA, as it was used for third-generation legacy applications. The fact that fourth-generation languages are less procedural than third-generation languages greatly assists this transition. Fourth-generation language systems, by simply looking at the SQL SELECT statements can identify which data files are used by the application. Using logic modularity rules, an analyst can establish cohesive classes based on applications that use the same data. This can be accomplished without using DFDs, although reengineering using DFDs is always a more thorough method for analysts to follow.

Linkage of fourth-generation language legacy and packaged software applications needs to be accomplished after application reengineering is completed. As with third-generation language systems, this can either be accomplished using a data parameter or the creation of a special database. However, with fourth-generation languages it is likely that application integration will occur using databases, since both systems use them in their native architectures. An analyst will most likely find that application communication with fourth-generation languages will not always require separate databases to be designed solely for the purpose of system linkage. The more attractive solution to integration is to identify the data

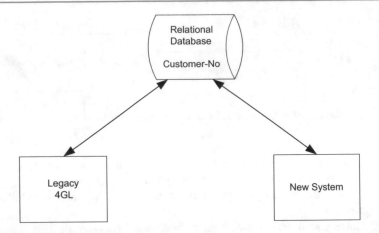

Fig. 10.13 Fourth-generation language legacy shared database architecture

elements that are common between the two systems so it can be shared in a central database available to all applications as shown in Fig. 10.13.

The use of CRUD in fourth-generation languages is used less, but is certainly applicable and should be implemented by the analyst if he/she feels that the code is too procedural. In other words, the code architecture resembles third-generation as opposed to fourth-generation.

10.13   "Leaving As Is"—Fourth-Generation Legacy Systems

The process of limiting integration to just the sharing of data is similar to the design architecture that I used for third-generation language systems. Indeed, the architecture of linking separate and distinct software systems can only be accomplished by sharing common data. Once again, this data can be shared either using a data parameter or data file.

Because many fourth-generation language systems utilize the same architecture as a packaged software system (three-tier client/server using Windows NTor UNIX/LINUX), it is sometimes advantageous to make use of certain operating system level communication facilities. For example, UNIX allows applications to pass data using an operating system facility called a "pipe." A pipe resembles a parameter, in that it allows an application to pass a message or data to another application without creating an actual new data structure, like a database. Furthermore, a pipe uses an access method called "FIFO" (first-in, first-out), which is the same access criteria used by parameters. FIFO also requires that once the data is read, it cannot be read again. The major advantage of using a pipe is that the message/data can be stored long after the application that created the message has terminated in memory. Thus, linkage of information among IoT, packaged

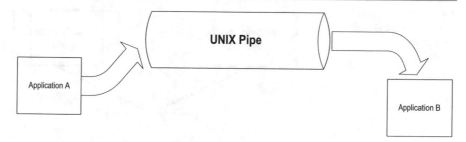

Fig. 10.14 Intra-application communication using a UNIX pipe

software, and fourth-generation language applications can be accomplished in RAM at execution time, which is called "intra-application communication." This capability reduces overhead as well as the need for separate modules to be designed that would just handle data communication as shown in Fig. 10.14.

10.14 Hybrid Methods: The Gateway Approach

Thus far in this chapter I have focused on the interface between a specific type of legacy and IoT, blockchain and/or packaged software systems. Each type was defined with respect to its "generation" type. In reality, however, legacy systems are not that self-defined. Many large organizations have "legacy layers," meaning that multiple generations exist throughout the enterprise. In this case, attempting to integrate each generation with a central packaged software system is difficult and time consuming. Indeed, migrating and integrating legacy systems is difficult enough. In these complex models, another method used for migration of legacy applications is a "hybrid" approach called "Gateway." The gateway approach means that there will be a software module that mediates requests between the packaged software system and the legacy applications. In many ways, a gateway performs similar tasks as a TP system. Thus, the gateway acts as a broker between applications. Specifically, gateways:

- Separate yet integrate components from different generation languages. It allows for the linkages among multiple generation language systems.
- Translates requests and data between multiple components.
- Coordinates between multiple components to ensure update consistency. This means that the gateway will ensure that redundant data elements are synchronized.

Typical gateway architectures would be designed as shown in Fig. 10.15.

Fig. 10.15 Gateway
architecture for legacy
integration

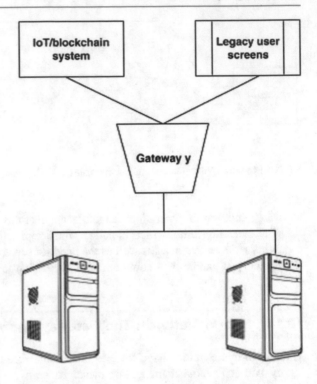

The most beneficial role of the gateway is that it allows for the phasing of legacy components. The infrastructure provides for an incremental approach to conversion by establishing a consistent update process for both data and applications.

10.15   Incremental Application Integration

A gateway establishes a transparency for graphical user interfaces (GUI), character-based interfaces, and automated interfaces (batch updates) to appear the same to the packaged software system. Hence, the gateway insulates the legacy system so that its interface with the packaged software systems seems seamless. This is accomplished through an interface that translates requests for process functions and routes them to their appropriate application, regardless of the generation of the software and its particular phase in the packaged software migration. Figure 10.16 depicts the process functions of the gateway system.

The most salient benefit of the gateway approach is its consistency with the object-oriented paradigm and the concept of application reusability. Specifically, it allows any module to behave "like" a reusable component notwithstanding its technical design. Under this architectural philosophy, a particular program, let's say, a third-generation language system, may eventually be replaced and placed into the gateway, with temporary bridges built until the overall migration is completed.

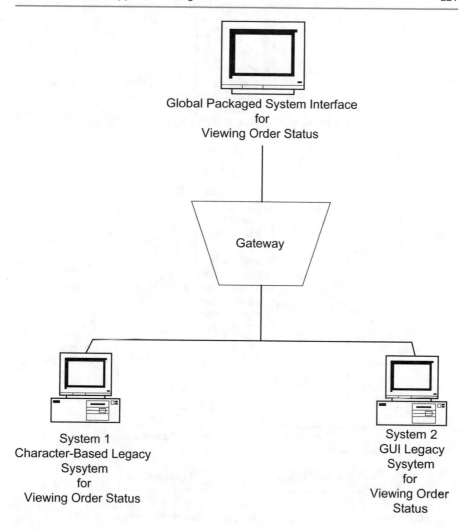

Fig. 10.16 Application functions of legacy gateways

This procedure also supports a more "global" view of the enterprise as opposed to just focusing on a particular subsystem. Figure 10.17 depicts the concept of process integration using the gateway architecture.

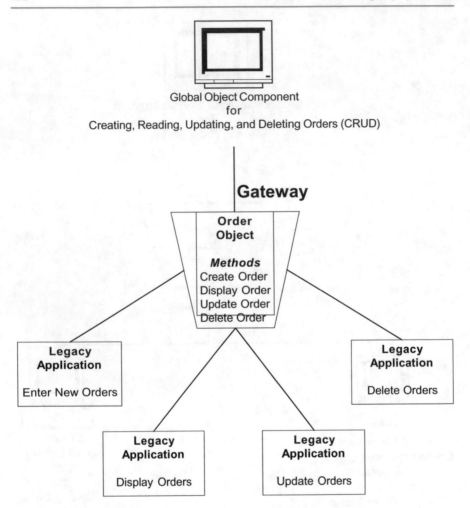

Fig. 10.17 Process integration and migration using gateway architecture

10.16   Incremental Data Integration

Incremental data integration focuses on the challenge of keeping multiple sets of data coordinated throughout a packaged software system.

The two primary issues relating to data integration focus on queries and updates. Queries involve the access of complete information about a dataset (collection of related data elements) across multiple systems. Much of the query challenges can be addressed by using a data warehouse or data mining architecture. The gateway would serve as the infrastructure that would determine how many copies of the data exist and its location.

The more difficult and more important concept of data integration is the ability of the gateway to coordinate multiple updates across databases and flat-file systems. This means that the changing of a data element in one component would "trigger" an automatic update to the other components. There are four scenarios that could exist regarding the different definitions of data elements:

1. The data elements in each system have the same name. This at least allows analysts to identify how many copies of the element exist in the system.
2. The data elements do not match-up by name. This requires that the analyst design a "mapping" algorithm that tracks the corresponding name of each alias.
3. Data elements match by name but not by attribute. In this case the analyst must propagate updates to the data element by tracking the different attribute definitions it has across systems. These differences can vary dramatically. The most obvious is element length. If the length of the data element is shorter than the one that has been updated then there is the problem of field truncation. This means that either the beginning value of ending value of the string will be lost when the value is propagated to the system with the shorter length definition. On the other hand, if the target is longer, then the process must populate either the beginning or end of the string so that the element has a complete value. This is graphically depicted in Fig. 10.18.
   Furthermore, the same data element might have different data types, meaning that one is alphanumeric and the other numeric. In this case, analysts need to know that certain values (e.g., a leading zero) will not be stored in the same way depending on its data type classification.

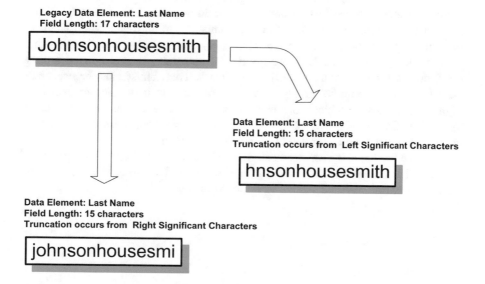

Legacy Data Element: Last Name
Field Length: 17 characters

Johnsonhousesmith

Data Element: Last Name
Field Length: 15 characters
Truncation occurs from Left Significant Characters

hnsonhousesmith

Data Element: Last Name
Field Length: 15 characters
Truncation occurs from Right Significant Characters

johnsonhousesmi

Fig. 10.18 Propagating data elements with different field lengths

4. There is not a one-to-one relationship among data elements. This suggests that a data element in one system may be based on the results of a calculation (a derived data element). This would require a more in-depth analysis and mapping often solved by creating a stored-procedure that replicates the business rule to calculate the data element's value. So in this case there might be simple copies of the element moved from one system to another, as well as one data element value that needs to first be calculated and then propagated across multiple systems. For example, if the data element "Total-Amount" is entered in one system but calculated as Quantity times Price in another, the propagation of the values is very complex. First, the analyst must know whether the calculated value is performed first before the resultant value, in this case, Total-Amount is reentered in another system. If this is true, then the propagation is much easier; once the calculation is made then the result is copied to the "entered" element. The converse is much more complex. If the Total-Amount was entered, but the values of Quantity and Price were not, then it would be very difficult to propagate until both Quantity and Price were entered. The example is further complicated if adjustments are made to the Quantity, Price, or Total-Amount. For any change, the systems would need to automatically be "triggered" to recalculate the values to ensure they are in synchronization. Figure 10.19 graphically shows this process.

10.17 Converting Legacy Character-Based Screens

It would be naïve to assume that most legacy systems do not have character-based screens. Character-based screens are those that do not make use of the GUI. While most character-based screens in existence emanate from third-generation language mainframe implementations there are also many early fourth-generation language systems that preceded the GUI paradigm. Unfortunately, character-based screens often do not map easily to its GUI counterparts. The analyst must be especially careful not to attempt to simply duplicate the screens in the legacy software. Figure 10.20 shows a typical character-based legacy screen. Note that there can be up to four Contract/PO's as shown in the upper right-hand corner. The user is required to enter each Contract/PO on a separate screen.

On the other hand, the replacement GUI screen in Fig. 10.21 takes advantage of the view bar that allows for scrolling in a window. Therefore, the GUI version requires only one physical screen, as opposed to four.

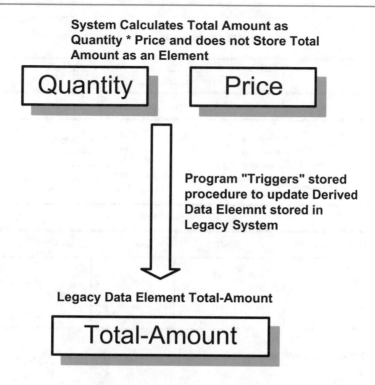

System Calculates Total Amount as Quantity * Price and does not Store Total Amount as an Element

Quantity   **Price**

Program "Triggers" stored procedure to update Derived Data Eleemnt stored in Legacy System

Legacy Data Element Total-Amount

Total-Amount

Fig. 10.19 Propagation of calculated data elements

10.18   The Challenge with Encoded Legacy Screen Values

In most legacy character-based screens, a common practice was used to create codes that represented another, more meaningful data value. For example, numeric codes (1, 2, 3, 4, etc.) might be used to represent product colors such as blue, green, dark red, etc. Legacy applications used codes because they reduced the number of characters needed to type in the value on a screen. The technology to implement common GUI features such as drop-down menus and pop-up windows were not available. Indeed, many people used codes just from habit, or had to use them in order to implement computer systems. When transitioning to a GUI system, especially on the Web, it is wise to phase-out any data elements that are in an encoded form unless the codes are user-defined and are meaningful within the industry or business area. This essentially means that certain codes, like State (NY, CT, CA, etc.) are industry standards that are required by the business, as opposed to those created to aid in the implementation of software—like color codes. In the later case, the color name itself is unique and would be stored in an entity with just its

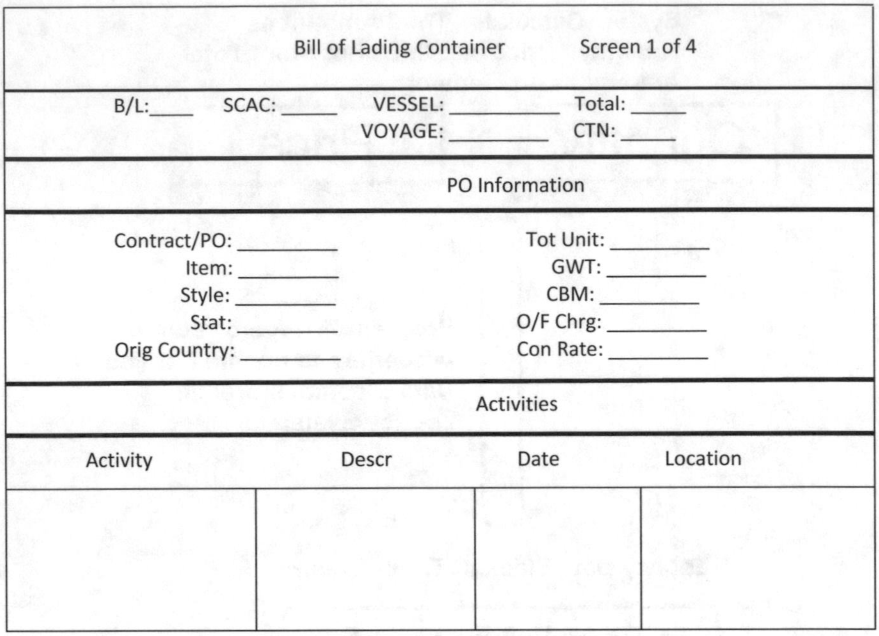

Fig. 10.20 Character-based user screen

descriptive name, as opposed to a code, which then identifies the actual description. Figure 10.22 shows the character-based and GUI screen transition.

Changing character-based screens that contain encoded values has a trickle-down effect on the data dictionary and then on logic data modeling. First, the elimination of a coded value inevitably deletes a data element from the data dictionary. Second, codes are often key attributes, which become primary keys of entities. The elimination of the code, therefore, will eliminate the primary key of the entity. The new primary-key will likely be the element name. These changes must then be made to the entity relational diagram (ERD) and placed in production (see Fig. 10.23).

Third, the elimination of codes affects previous stored-procedures that use queries against the coded vales. Therefore, analysts must be sure to reengineer all queries that use the codes. This transition will add tremendous value since encoded elements typically add unnecessary overhead and time delays to queries. Finally, the elimination of encoded values will free up considerable space and index overhead. This will result in an increase in performance of the legacy system.

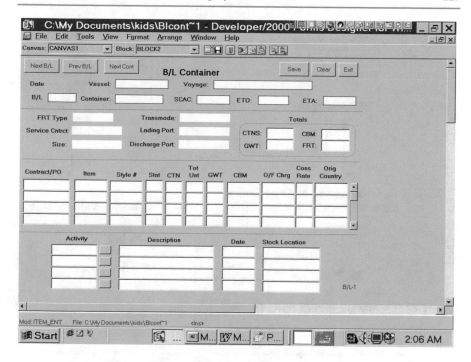

Fig. 10.21   Transformed character-based to GUI screen

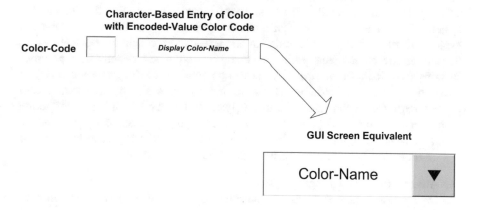

Fig. 10.22   Encoded value GUI screen transition

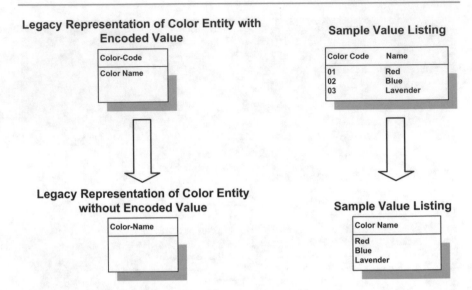

Fig. 10.23 Transition to encoded databases

10.19   Legacy Migration Methodology

As stated earlier, all legacy systems inevitably must reach the end of their original life cycle. Therefore, notwithstanding whether certain components will remain "as is" or enhanced, eventually IT management must plan for migration to another system. The issue that this section addresses is how to establish a migration life cycle that takes into consideration an incremental approach to replacement of various legacy components within an enterprise computer system. The previous sections provided a framework of what can be done with legacy systems and their integration with other systems. This section provides a step-by-step template of procedures to follow that can assist the analysts on the schedule of legacy migration including temporary and permanent integration. This approach is an incremental one, so analysts can use it as a checklist of the progression they have made in the legacy migration life cycle. In all there are 12 steps as follows:

1. Analyze the existing legacy systems.
2. Decompose legacy systems to determine schedules of migration and linkage strategies.
3. Design "As Is" links.
4. Design legacy enhancements.
5. Design legacy replacements.
6. Design and integrate new databases.
7. Determine new infrastructure and environment, including gateways.

8. Implement enhancements.
9. Implement links.
10. Migrate legacy databases.
11. Migrate replacement legacy applications.
12. Incrementally cutover to new systems.

The above steps are graphically depicted in Fig. 10.24.

Note that there are two streams of steps, that is, steps 3–5 and 8–10 that can occur in parallel. These steps encompass the three types of legacy migration choices that can be made: replacement, enhancement, and "as is." While this life cycle seems simple, in reality it is a significant challenge for most migrations to plan, manage, and modify these steps and their interactions. Indeed, creating a migration plan and adequately coordinating the incremental and parallel steps is a difficult project. The subsequent sections will provide more details for each of these 12 steps.

Step 1: Analyze the Existing Legacy Systems

It is obviously important that analysts fully understand all of the existing legacy components that exist in the system. The objective is to provide the requirements of each system and how it relates to the system. Analysts must remember that little to no documentation will be available to fully represent the architecture of the legacy system. However, analysts should compile as much information that is available including but not limited to:

• User and programming documentation
• Existing users, software developers and managers
• Required inputs and outputs and known services of the legacy system
• Any historical perspective on the history of the system itself.

Regardless of what existing information and documentation is available, certain aspects of reverse engineering must be used. Various CASE (Computer Aided Software Engineering) tools should be used that allows analysts to create a repository of data and certain levels of code analysis, particularly for third-generation language migrations. The analyst should create DFDs and PFDs for process analysis, and logic data modeling (LDM) and entity relational dia-gramming (ERD) for representation of the data. The analyst should also determine which legacy components are decomposable and non-decomposable. Inevitably, regardless of whether the decision is to replace immediately, enhance, or leave "as is," ultimately little of the existing code will survive the ultimate migration of a legacy system (Brodie and Stonebraker 1995).

Legacy Migration Life Cycle

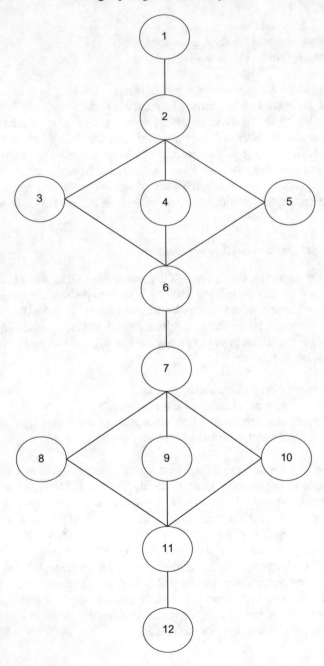

Fig. 10.24 Legacy migration life cycle

Step 2: Decompose Legacy Systems to Determine Schedules of Migration and Linkage Strategies

The gradual migration of legacy systems is easiest accomplished when analysts utilize decomposition engineering. Previous chapters have outlined the process of functional decomposition, which is based on the breaking down of a system into its functional components. Decomposition, which results in components also allows for reusability of code. Since the fundamental premises of a package system are reusability, the process of decomposition is a mandatory part of the life cycle of any legacy migration. Thus, analysts should decompose all DFDs to functional primitives. Process analysis continues with the writing of process specifications for each functional primitive. These functional primitives will either be rewritten from the existing code or documentation or recreated from analyzing the functionality of the program. Analysts need to remove all dependency logic that links modules from the legacy code because it represents coupling among the programs. Module dependencies can typically be identified by finding *procedure calls* from within the legacy code. Ultimately, each of these process specifications will become methods. Eventually all methods will be mapped to classes and identify the attributes that each class needs. While this sounds simple, there will be a number of processes for which decomposition will be problematic. This will typically occur for legacy code that is too eclectic and needs to simply be reengineered. In these cases, analysts may want to interview users to understand the functionality that is required, as opposed to relying solely on the written legacy code.

From a data perspective, entity relational diagrams (ERD) need to undergo normalization. Remember that the DD and ERD produced in Step 1 is a mirror of the existing system, which most likely will not be in third-normal form. Thus this step will result in the propagation of new entities and even new data elements. Furthermore, data redundancies will be discovered as well as derived elements that should be removed from the logic model. However, while these steps are taking place, analysts need to be cognizant that the process of normalization represents the total new blueprint of how the data should have been originally engineered. The actual removal of elements or reconstruction of the physical database needs to be a phased-plan in accordance with the overall legacy migration effort. Thus Step 2 provides the decomposed framework, but not the schedule of implementation.

Finally, analysts must always be aware of the challenges of decomposition, particularly over-decomposition, meaning that too many classes have been formed that will ultimately hurt system performance. There needs to be a mix of decomposed levels, which will serve as the basis of the new migration architecture.

Step 3: Design "As Is" Links

This step involves determining and designing what components will remain untouched except for linkages that are necessary with other package software components. These modules are determined not to be part of the initial migration plans; however, they need to function *within* the packaged software system infrastructure and are therefore part of its architecture. Part of the decision needs to include how data will be migrated into the packaged software framework. In most cases, "as is" components continue to use their legacy data sources. Consideration must be given to how legacy data will be communicated to other components in the packaged software system. Analysts need to consider either a parameter-based communication system, or a centrally shared database repository as outlined earlier.

Step 4: Design Legacy Enhancements

This step determines which modules will be enhanced. BPR (business process reengineering) will be used to design new features and functions in each business area. Analysts should identify essential components and determine what changes need to be employed to make the existing system behave more like an object-oriented system. Common processes and databases also need to be mapped so that shared resources can be designed between legacy and packaged software systems. New linkages will also be needed and the analyst must determine whether to use parameters or databases, or both to implement the communication among application systems.

User screens may also need to be updated as necessary, especially to remove encoded values or moving certain character-based screens to GUI. Many of the enhancements to a legacy application are implemented based on the analysis performed in Step 1. Any modifications will need to eventually operate on newer platforms once the total migration of legacy systems is completed. Analysts need to be cognizant that additional requirements mean increasing risk, which should be avoided when possible. However, during enhancement consideration it is almost impossible to ignore new requirements. Therefore, analysts need to focus on risk assessment as part of the life cycle of legacy migration.

Step 5: Design Legacy Replacements.

Analysts must focus on how to reconstruct logic in a later generation of software architecture. Therefore, it is important to understand the differences in generation language design. This chapter provided two types of legacy software systems: third- and fourth-generation languages. Third-generation languages were depicted as being more procedural and more difficult to convert.

Analysts must design the target applications so they will operate in accordance with the business rules and processes that will be supported in the new systems environment. Because of this integration, most replacement legacy migrations require significant reengineering activities. These activities necessitate the inclusion of new business rules that may have evolved since the legacy system was placed in production. Furthermore, new business rules can be created simply by the requirement of being an packaged software system.

Another important component of legacy migration is screen design. When replacing legacy systems, analysts must view the migration as assimilation, that is, the old system is becoming part of the new one. As a result, all existing packaged software screens need to be reviewed so that designers can determine whether they need to be modified to adopt some of the legacy functionality. This is not to suggest that all of the legacy systems screens will be absorbed into the existing packaged software system. Rather, there will be a combination of new and absorbed functionality added to the target environment.

Step 6: Design and Integrate New Databases

From an enterprise perspective, analysts must gather all of the permutations of legacy systems that exist and seek to provide a plan on how to integrate the data into one source. For "as is" solutions, legacy data files will most likely remain separate from the packaged software system until the complete legacy is migrated. However, the process of enhancing and replacing legacy data should have the objective of creating one central database source that will serve the entire packaged software enterprise system. Indeed, a central data source reduces data redundancy and significantly increases data integrity.

The process of data integration can only be accomplished by "combining user views" which is the process of matching up the multiple data element definitions that overlap and identifying data redundancies and alternative definitions. This can only be completed by creating a repository of data elements and by representing the data graphically using an ERD. Once each system is represented in this fashion then the analyst must perform logic data modeling as prescribed in Chap. 9. The result will be the creation of one central database system cluster that can provide the type of integration necessary for successful packaged software implementations.

While this is the goal for all analysts, the road to successful implementation is challenging. First, the process of normalization will require that some data elements be deleted (e.g., encoded values) while others will need to be added. Second, full data integration cannot be attained until all replacement screens are complete. Third, applications must be redesigned so that the centrality of the data source is assumed to exist. Since this process may be very time-consuming, it may not be feasible to attempt a full database migration at one time. Therefore, the legacy data may need to be logically partitioned to facilitate incremental database migration. Thus, there will be legacy data subsets that are created to remain independent of the central database until some later migration phase is deemed feasible. Of course, this strategy requires the design of temporary "bridges" that allows the entire packaged software system to appear cohesive to users.

Another important factor in planning data migration is to determine how much is really known about the legacy data itself. The less knowledge available, the longer the period where legacy data and packaged software data need to run separately and in parallel.

Step 7: Determine New Infrastructure and Environment, including Gateways

Prior to the migration of any system, the necessary hardware and software infrastructure must be planned and installed. A common error in legacy migration is not factoring the time and effort to provide this infrastructure. Furthermore, this process, which may create a new network environment, needs to determine the placement of software in a three-tier client/server platform. This means that further decomposition may be required of all processes, especially object classes. You may recall in Chap. 8 that classes may undergo further decomposition depending on the need to distribute application across the network.

Another important factor is performance. In many instances it is difficult for network engineers to predict performance in large application system environments. It may be necessary to plan for several benchmarks in performance early in the design phase. Benchmarking is the process of setting up an environment that replicates a production network so that modifications can be made, if necessary, to the design of the system to increase performance.

Another decision that must be made at this juncture is whether a gateway infrastructure will be created to mediate legacy layers. This decision, of course is highly dependent on the migration life cycle; the more legacy layers that will be phased in over time, the higher the chances that a gateway processor for data and applications will be necessary. The decision to go with a gateway is significant, not only from the perspective of software design, but network infrastructure as well. Constructing a gateway can be very costly. It involves writing the system from scratch or tailoring a commercial product to meet the migration requirements. It is also costly because of the amount of additional hardware necessary to optimize the performance of the gateway servers. However, the benefits of a gateway are real, as it could provide a dependable structure to slowly migrate all components in the new systems environment.

Step 8: Implement Enhancements

This step requires a schedule of when legacy enhancements will be implemented and become part of the production system. Many analysts suggest that the simplest modules go into production first so that any unexpected problems can be dealt with quickly and efficiently. Furthermore, simple modules tend to have small consequences should there be a problem in processing or performance. There are some other aspects of coordination, however. For example enhancements that feed off of the same data or use the same subsystems should obviously be implemented at one time.

Another factor in the decision of which enhanced components go first relates to the affects it has on other subsystems. This means that the priority may indeed be influenced by what other systems need or are dependent upon from other systems. Another issue could be the nature of the legacy links. Should a link be very complex or dependent on other subsystem enhancements, its schedule could be affected. Finally, the nature of the enhancements has much to do with the decision as opposed to just the application's complexity. There may be simple enhancements that are crucial for the packaged software system and vice versa.

Step 9: Implement Links

As I alluded to in Step 8, the determination of legacy links greatly affects the scheduling of the migration cycle. Once the determination is made in Step 8, the related legacy links must be put in place. This could also mean that the gateway, if designed, must also be in operation since many links might be filtered through the gateway infrastructure. So, the implementation of legacy links relates to both hardware and software. Notwithstanding whether a gateway is in place, database links often require separate servers. In many cases because these "servers" interface with the internet, there is a need to install firewalls to ensure security protection.

From a software perspective, legacy links can almost be treated like conversion programs. There needs to be substantial testing done to ensure they work properly. Once legacy links are in production, like data conversions, they tend to keep working. It is also important to ensure that legacy links are documented. Indeed, any link will eventually be changed based on the incremental migration schedule. Remember that most legacy links are accomplished by building temporary "bridges." The concept of temporary can be dangerous, in that many of these links, over time, tend to be more permanent. That is, their temporary life can sometimes extend beyond the predicted life of a permanent component. The message here should be that legacy links, while they are a temporary solution, should be designed under the same intensity and adherence to quality as any other software development component.

Step 10: Migrate Legacy Databases

The migration of data is so complex that it should be handled as a separate and distinct step in the migration life cycle. Data affects everything in the system, and often if it is not migrated properly it can cause immense problems. First, the analyst must decide on the phasing of data based on the schedule of application migration. Hopefully, the process of data migration should be done in parallel to Steps 8 and 9.

The most challenging aspect of data migration is the physical steps in the process. Migrating new entities and schema changes are complex. For example, changes to databases require that the tables be "dropped" meaning that they are taken off-line. Data dictionaries need to be updated, and changes to stored procedures and triggers are extremely time-consuming. Most problematic is the process f quality assurance. While some testing can be done in a controlled environment, most of the final testing must be done once the system is actually in production. Therefore, the coordination with users to test the system early is critical. Furthermore, there must be back-up procedures in case the database migration does not work properly. This means that there is an alternate fail-safe plan to reinstall the old system should major problems arise. Finally, a programming team should be ready to deal with any problems that arise that do not warrant reinstalling the old version.

This might include the discovery of application "bugs" that can be fixed within a reasonable period and are not deemed critical to operations (which means there is usually a "work-around" for the problem). Analysts must understand that this process must be followed each time a new database migration takes place!

Database migration is even more complex when there is a gateway. The reason is that the gateway, from an incremental perspective, contains more and more database responsibilities each time there is a migration. Therefore, for every migration, the amount of data that can be potentially affected grows larger. In addition, the number of data that becomes integrated usually grows exponentially, so the planning and conversion process becomes a critical path to successful migration life cycles. Since the migration of legacy databases becomes so much more difficult as the project progresses, the end of the life cycle becomes even more challenging to reach. That is why many migrations have never been completed!

Step 11: Migrate Replacement Legacy Applications

Once the database migration is completed, then the remainder of the legacy applications can be migrated to the new system. These applications are usually the replacement components, which have been reengineered in the object-oriented paradigm. These programs, then, have been designed to operate against the target databases with the new functionality required for the packaged software system. Since replacement applications usually do not create links, there is typically little effect on gateway operations. What is more challenging is the quality assurance process. Users need to be aware that the code is relatively new and will contain problems regardless of the amount of pre-production testing that has been performed. In any event, programmers, database administrators, and quality assurance personnel should be on-call for weeks after system cutover.

Step 12: Incrementally Cutover to New Systems

As discussed above, testing and application turnover are two areas that frequently are overlooked. Because projects typically run over budget and schedule, the final procedures like testing and verification are usually shortened. The results of this decision can be devastating to successful legacy migrations. Because of the size and complexity of many packaged software systems, to go "cold turkey" is unrealistic and irresponsible. Therefore, an analyst should consider providing test scenarios that provide more confidence that the system is ready to be cutover. This approach is called "acceptance testing" and requires that users be involved in the determination of what tests must be performed before the system is ready to go live. Thus, acceptance test plans can be defined as the set of tests that if passed will establish that the software can be used in production. Acceptance tests need to be established early in the product life cycle and should begin during the analysis phase. It is only logical then that the development of acceptance test plans should involve analysts. As with requirements development, the analyst must participate with the user community. Only users can make the final decision about the content and scope of

Quality Assurance

Acceptance Test Plan

Number:

Date:

Product: Contact - Using Enter Key

Test Plan #: 1G

Vendor:

QA Technician:

Test No.	Condition Being Tested	Expected Results	Actual Results	Comply Y/N	Comments
1	Enter LAST NAME for a new contact, press enter key. Repeat and enter FIRST NAME, press enter key	Should accept and prompt for COMPANY SITE			
2	Select COMPANY Site from picklist	Should accept and prompt for next field			
3	Enter LAST NAMEand FIRST NAME for a CONTACT that is already in the System.	Should accept and prompt for COMPANY SITE			

Fig. 10.25 Acceptance test plan

the test plans. The design and development of acceptance test plans should not be confused with the testing phase of the software development life cycle.

Another perspective on acceptance testing is that it becomes a formal checklist that defines the minimal criteria for incrementally migrating systems. However, one must work with the understanding that no new product will ever be fault-free. The permutations of testing everything would make the timetable for completion unacceptable and cost prohibitive. Therefore, the acceptance test plan is a strategy to get the most important components tested completely enough for production. Figure 10.25 represents a sample acceptance test plan.

10.20 Problems and Exercises

1. What is a Legacy? Explain.
2. Describe the Five generation languages. What increases with each generation?
3. What are Essential Components?
4. What is the Object-Oriented paradigm mean?
5. How does Object Orientation relate to Business Area Analysis?
6. What is a Legacy Link?
7. Explain the notion of Logic Reconstruction.
8. What is a UNIX pipe?
9. Explain how Legacy Integration operates through Gateway Architecture.
10. What is Propagation?
11. What are the essential differences between character-based screens and GUI?
12. What is an encoded value?
13. What is the relationship between an object and an API?
14. What are the restrictions of CRUD in blockchain architecture? Why is this difference so important in a ledger-based system?

References

Brodie, M. L., & Stonebreaker, M. (1995). *Migrating legacy systems: Gateway interfaces and the incremental approach*. San Francisco, CA: Morgan Kaufmann Publisher Inc.
Stair, R. M., & Reynolds, C. W. (1999). *Principles of information systems* (4th ed.). Cambridge, MA: Course Technology.
Whitten, J., Bentley, L., & Barlow, V. (2000). *Systems Analysis & Design Methods* (3rd ed., p. 238).

Build Versus Buy

11

11.1 Overview

This chapter addresses a difficult and controversial decision that is made every time an organization seeks a software solution to meet its needs: do we make it to our specific needs, or do we buy something that is made to order but may not do everything we want? Often the build decision is called the "make" alternative and suggests that the product will be made in-house versus the buy concept that can be referred to as outsourcing. I do not believe these simple labels are accurate or appropriate. Whether something is built or bought has little to do with whether the process is outsourced, so we need to be careful in the way we label these two alternatives.

Inman et al. (2011) suggests that Build and Buy decisions need to be made at both the strategic and operational levels. Burt et al. (2003) provided some direction on the strategic reasons for buying and tied it into a definition of outsourcing, providing three concrete categories of reasons NOT to outsource:

1. The item is critical to the success of the overall product and is perceived so by the company's clientele
2. The item requires specialized design and skills and such skills are limited in the organization
3. The item fits into the firm's core competencies but needs to be developed in the future.

Historically, most organizations selected the buy option because their assumption was that it lowered costs. Indeed, over 70% of product ownership usually occurs after implementation. So where is this cost? It occurs in maintenance, where in-house teams must continually alter and develop. The off-the-shelf concept tends to keep costs lower because in theory the sum of all the clients will create better software for all. Others suggest that a company should always first seek a package because of common business challenges including:

© Springer Nature Switzerland AG 2020
A. M. Langer, *Analysis and Design of Next-Generation Software Architectures*,
https://doi.org/10.1007/978-3-030-36899-9_11

- Cost
- Time to market
- Political situation in the environment
- Architectural differences
- Skill sets of existing staff.

However, The Gartner Group in 2003 published a report that suggested this trend was changing and that there were a growing number of firms that were returning to building applications internally. Gartner cited the following reasons for the change of heart:

- Rising competitive advantages of using emerging technologies
- Increased availability of talented software developers
- Poor reputation of prior uses of package software—that it is not agile enough and difficult for departments to use
- Increased needs to adapt to unique and changing business needs.

Ledeen's (2009) and Moore's (2002) analysis of how to approach make versus buy is quite useful. He established a step-by-step criterion to help organizations to make the best decision. This criterion included the following:

- Core versus Context
- Coverage
- Direction
- TCO
- Scale
- Timing
- Standards.

11.2   Core Versus Context

This decision point relates to the strategic importance of the application. The more strategic the application, the more likely that the organization develop software internally (also see Langer 2011, below). The concept is simple: if the application relates to basic functions in accounting, HR or payroll, then it is not core. However, software used by WalMart, although accounting related is used as a supply chain management that drives every aspect of their competitive advantage as a company. The result of course is that Walmart developed their supply chain as a core and unique application. Moore's chart (Table 11.1) provides an interesting matrix of how core can be determined.

The chart above reflects that Mission Critical application should be developed in-house with Context oriented applications may be modified to meet package requirements.

Table 11.1 Build versus buy chart		Core Engage	Context Disengage
	Mission critical (control)	MAKE	OUTSOURCE
	Supporting (entrust)	PARTNER	CONTRACT

11.3 Coverage

The coverage assesses the extent of the match of the packaged product with the business requirements. The general rule is that a package should have at least 80% of the features and functions needed by the organization. However, Ledeen states that this could be a trap, suggesting that a package's capabilities outside of the immediate needs of the business are equally as important. This is relevant, given that business needs are continually evolving, organizations must be cognizant of having applications that can not only handle what is, but what may be future needs of the business. In addition, a certain feature in a packaged solution may actually offer a better business alternative than currently used by the business—so it's a complex issue as they say.

11.4 Direction

The key words relating to Direction is flexibility, maintainability, and extendibility of the software throughout its life. Ultimately, Direction relates to how much control the organization has over the product, especially those products that may need to change—that is, the volatility of what the software does. For example, if the product is a basic accounting system, it is likely not to change substantially over its life. However, if it is a healthcare product that it regulated by government requirements in a highly fragile market, then Direction is a key decision factor. Much of this relates to the design and architecture of the product itself. Can it be easily modified? What is the extent of changes that can be controlled via user input? All of these factors are significant to whether a package is the wise choice.

11.5 Total Cost of Ownership (TCO)

The TCO represents to entire cost. Components of this cost include license fees for the product, maintenance, product customizations, and support. The major variable in TCO is custom modifications. Vendors will often provide an estimate but managers need to be careful of "scope-creep" where original requirements for customizations become greatly expanded during the design phase of the project. A good approach that can help this decision is to determine the number of features and functions in the package that the organization does *not* need. A packaged solution that has many features and functions that are not needed might suggest that the application is not a great match and is likely designed for a different audience. It is important to note that many application packages were developed first as a custom application for a specific client and then tailored for others. This process was a typical evolution of how many software packages evolved in the market. Therefore, understanding the history of how the application package was developed might provide a hint as to the TOC and the fit in general—the two tend to go together.

11.6   Scale

The size of the package is a factor, especially when it has many modules. This is very relevant in large Enterprise Resource Planning (ERP) products where a high degree of scaled interoperability is important. These modular products also allow clients to purchases business components at later periods and easily retrofit them. However, if it is *not* the intention of the organization to scale, then a large integrated package may be overkill.

11.7   Timing

There are many who would believe that a packaged solution will be faster to implement—but be careful-often this is not the case. Packaged solutions may add steps to the SDLC and as a result could actually prolong the timing of going Live on the product. Ledeen suggests that while COTS provides greater predictability it could be a reflection on the limits of flexibility and imposed restrictions. Either case I would strongly recommend that the decision whether to have a make versus buy not be contended solely based on speed to completion—the organization may be in for a big surprise. The more an organization can accept the base package the faster the implementation will be, for sure. However, accepting the package as is does not necessarily mean that it is the best thing for the business.

11.8   Standards

Ledeen defines standards as the consistency across systems of the way things get done. I see this issue more in terms of consistency of architecture of the organization. This means that the hardware platforms and software architecture (middleware, office products, etc.) should be consistent for packaged software to maximize its benefits. If this is not the case, that is, where you have multiple architectures across the organization, then the value of a package become less evident. This is especially true if the package requires a specific architecture for all of the systems. This is why open systems are so attractive to organizations Unfortunately in multi-national firms, having many architectures is not unusual. Much of this may have occurred due to corporate acquisitions of other firms—when you acquire a new business you often acquire a new systems architecture—both hardware and software!

11.9 Other Evaluation Criteria

The above issues are important but there are other contextual things to consider when making the decision:

- *Complexity of Product*: the more intricate the software application, the harder for COTS to work. Complex products also tend to shorten the life cycle and have more evolving needs.
- *State of the Art*: Users who seek packaged solutions often require the latest and greatest up-to-date product. They may be disappointed to learn that package software has some unique limitations in this area. First, vendors of packages have user bases to take care of—who ultimately have older versions and hardware that they must support. This results in difficulties with downward compatibilities of their user base. Just look at Microsoft's problems when they upgrade operating systems and software. IBM used to issue "no support" edicts to its customers who failed to upgrade their products over time. The better vendors force their users to upgrade but it is always messy.
- *Maintenance*: COTS often involves product upgrades and new maintenance releases. Sometimes maintenance releases include bug fixes and workarounds for packaged discovered problems. Maintenance can also be very tricky—how do organizations load new versions from the vendors when there are customizations, for example? It gets to be a challenge especially when the updates are regulatory in nature. COTS that have customizations inevitably need to go through a retrofit, where the customized portions have to be "re-customized" to deal with packaged software upgrades. This certainly adds to the cost equation over the life of the package.

While the above suggestions offer guides on what to think about and measure, the process is still complex without a real scientific methodology to determine ultimately whether to make or buy or both. However, Langer (2011) provides yet another concept called Driver/Supporter theory in which "buy" decisions would only be made for applications that were considered "Supporters." Figure 11.1 depicts a supporter item as something that has reached the stage of commodity notated in the circle labeled as "Economies of Scale."

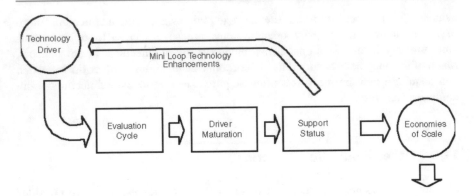

Fig. 11.1 Langer's Driver/Supporter life cycle

The diagram shows that all technology needs start out as Drivers but eventually become Supporters, thus lose their uniqueness in the marketplace. It is as if you were creating a strategic advantage by implementing a new email system—this would not really provide the organization with a competitive advantage—rather as an enabler to compete—a very different concept. The point I am making here is that a new email system would likely be "bought" and possibly implemented using an outsource vendor. In accordance with Burt et al. then this would be a case for buying as opposed to making.

11.10   Drivers and Supporters

I think this concept of Driver/Supporter is essential to understanding make/buy decisions. This section provides further details on this theory and practice as a vehicle for better determination of make versus buy.

To summarize Driver/Supporter, there are essentially two types of generic functions performed by departments in organizations: Driver functions and Supporter functions. These functions relate to the essential behaviour and nature of what a department contributes to the goals of the organization. I first encountered the concept of drivers and supporters at Coopers & Lybrand, which was at that time a Big 8[1] accounting firm. I studied the formulation of Driver versus Supporter as it related to the role of our EDP (Electronic Data Processing) department. The firm was attempting to categorize the EDP department as either a Driver or Supporter.

Drivers are defined as those units that engaged in front-line or direct revenue generating activities. Supporters are units that did not generate obvious direct revenues but, rather, were designed to support front-line activities. For example, operations such as internal accounting, purchasing, or office management were all classified as supporter departments. Supporter departments, due to their very nature, were evaluated on their effectiveness and efficiency or economies of scale. In contrast, driver organizations are expected to generate direct revenues and other ROI values for the firm. What was also interesting to me at the time was that Drivers were expected to be more daring—since they must inevitably generate returns for the business. As such, Drivers engage in what Bradley and Nolan (1998) coined "sense and respond" behaviours and activities. Let me explain.

Marketing departments often generate new business by investing or "sensing" an opportunity, quickly—because of competitive forces in the marketplace. Thus, they must sense an opportunity and be allowed to respond to it in timely fashion. The process of sensing opportunity and responding with competitive products or services is a stage in the cycle that organizations need to support. Failures in the cycles of sense and respond are expected. Take, for example, the launching of new fall television shows. Each of the major stations goes through a process of "sensing" what shows might be interesting to the viewing audience. They "respond" after research and review with a number of new shows. Inevitably, only a few of these selected shows are actually successful; some fail almost immediately. While relatively few shows succeed, the process is acceptable and is seen by management as the consequence of an appropriate set of steps for competing effectively—even though the percentage of successful new shows is very low. Therefore, it is safe to say that driver organizations are expected to engage in high-risk oriented operations, of which many will fail for the sake of creating ultimately successful products or services.

The preceding example raises two questions: (1) How does "sense and respond" relate to the world of information technology, and (2) why is it important? Information technology is unique in that it is both a Driver and a Supporter. The latter being the generally accepted norm in most firms. Indeed, most IT functions are established to support a myriad of internal functions such as:

- Accounting and finance
- Data-Center infrastructure (e-mail, desktop, etc.)
- Enterprise level application (ERP)
- Customer support (CRM)
- Web and e-commerce activities.

As one would expect, these IT functions are viewed as overhead related, as somewhat of a commodity, and, thus, constantly managed on an economy-of-scale basis—that is, how can we make this operation more efficient, with a particular focus on cost containment?

So, what then are IT Driver functions? By definition, they are those that engage in direct revenues and identifiable return-on-investment (ROI). How do we define such functions in IT, as most activities are sheltered under the umbrella of marketing organization domains? (Excluding, of course, software application development firms that engage in marketing for their actual application products.) I define IT Driver functions as those projects that, if delivered, would change the relationship between the organization and its customers, that is, those activities that directly affect the classic definition of a market: forces of supply and demand, which are governed by the customer (demand) and the vendor (supplier) relationship.

The conclusion of this section, therefore, is that no Driver application product should be implemented using complete outsourcing, rather made in-house and owned by the firm. This does not, however, suggest that certain services and components be subcontracted out as long as the ownership remains within the company.

11.11 The Supporter Side of Buying

Based on the definition of a Driver, the Supporter side may indeed represent the need to buy a packaged solution. Since Supporter functions are "operational" by definition, they are considered to be a commodity and thus able to be implemented using more standardized application software. Thus, all of the advantages of using packaged software apply. In addition there should be less need for customization. For example, think of the choice to build an email system—you would only build it in house if you required unique capabilities that provided a competitive advantage. That is, the email system would be a Driver application because it would change the relationship between the buyer and seller. This is exactly the situation that occurred with Walmart, where what would ordinarily have been considered a commodity accounting system became an application of great strategic advantage. On the other hand, an email system that could provide such advantage would be unlikely for most organizations, and therefore they would seek a product that does what most organizations need in any email system—as a supporter solution.

11.12 Open Source Paradigm

Open Source software can be defined as free source code developed among a community that believes strongly in a free software movement. Initial examples of successful open source products are Linux and Netscape Communicator. The Open Source movement is supported under the auspices of the Open Source Initiative (OSI) that was formed in 1998 to provide guidance and standards of application.

As I previously mentioned the evolution of open source as an alternative to developing software has grown enormously in the software industry. Open source can also represent an option with make versus buy. Choices of whether to make or buy do not necessarily need to be binary; that is, one or the other, but rather could end up as a hybrid decision. For example, an organization can develop its own application using open source within its application development strategy or it can license a third party product that also contains open source. Finally, packages may be licensed that can be bridged or integrated with various open source modules. In any case, open source broadens the range of choices when determining the best application solution.

Open source users must, however, agree to the following conditions of use as well as providing conditions of use to others:

- Free distribution.
- Inclusion of source code.
- License must allow modifications and derived works.
- Allowed redistribution of modifications under the same license of the original software. License may require the derived work to carry a different name or version to protect the integrity of the original author.
- No discriminations against any specific groups or fields of endeavors.
- License cannot be restrictive to any software and be technology platform neutral.

To a certain extent open source provides organizations with the option to use package software that is free to modify and then offer their changes back to those that need it in the user base—so it can be a forum where organizations can share needs. The negative aspect to sharing is if the modification contains proprietary algorithms that represent a competitive advantage for the firm. The software must also be hardware neutral which presents challenges for those applications that run on proprietary systems. Still, open source applications are growing in popularity especially as a cloud computing option.

Furthermore, open source might present some unexpected legal issues particularly as it relates to ownership of the software. Suppose you use an open source routine or module in your proprietary application and then the company is acquired by another entity. The question then is who owns the product? Legally the portion that is open source cannot be owned, which creates a dilemma that was likely not foreseen by the organization's IT management. This dilemma is particularly relevant to vendor software products.

11.13 Cloud Computing Options

As I have discussed in previous chapters, cloud computing is the ultimate server-based paradigm to support IoT and blockchain technologies. Simply put, the host (Cloud) has all of the hardware, software, services and databases to support your business or enterprise. The organization essentially has the terminals and printers to do the work. Figure 11.2 depicts a cloud high-level configuration.

Beyond the connectivity, cloud is really about reduction of cost and perhaps using products that are shared by others like IoT devices. This does not mean that cloud products cannot have proprietary applications, rather that they have the ability to mix and match what is available in the cloud to meet specific needs of the organization. Because many of these applications can be shared in the cloud, the cost of ownership is very much lowered. Perhaps the largest benefit for using a

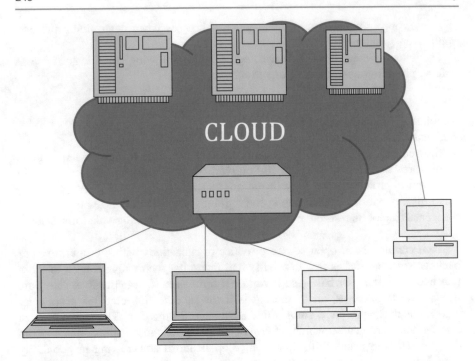

Fig. 11.2 Cloud configuration

cloud is in the savings for infrastructure and operations staff who would otherwise need to support the operation internally. We also know from Chap. 8 that cloud has a number of basic and complex configurations. It is worth reviewing these models to help determine the make versus buy decision.

11.14   Deployment Models

There are essentially five different deployable models for Cloud Computing:

1. *Public/External*: this is the basic model that allows users to access a network via the Internet and typically pay on a usage or application access basis. It resembles the 1970 concept of timesharing. *Clearly might be easier to buy.*
2. *Private/Internal*: a private cloud in many ways resembles an Intranet concept in that it is an internally developed shared service for the organization. As in an Intranet, a private cloud requires organizations to design the network and support it as if it were Public—of course with less complexity. *Likely a build choice.*
3. *Community*: this configuration represents a group of organizations that share resources. In effect it is a restricted public cloud—only certain organizations can

use it. Community clouds are attractive for specific industries that have similar needs or associations. *Could be both depending on the size of the shared community.*

4. *Hybrid*: a hybrid cloud really relates to providing specific administrative IT functions like backup, performance and security for both public and private cloud deployments. So, it is more a utility type of cloud service often provided by internal IT services or vendors like Oracle, etc. *Likely both with the private being developed and the public outsourced.*

5. *Combined*: This is the application of multiple types of clouds that allow organizations to enjoy the best provisions that it provides to each business. *By definition it can be either depending on the circumstances.*

Figure 11.3 depicts the graphical representations of these cloud type deployments.

Obviously, cloud computing has its drawbacks—it is essentially outsourcing major parts of your operation to a third party—for which is always a risk, so these should be measured as follows:

- *Security*: while all third parties promise security, history has shown that they can be pierced from the outside. So, security remains a concern for private and important data. For example, having healthcare information and other personal information in a Cloud could be very dangerous.
- *Governmental and political issues*: If stored data is kept in other locations under the auspices and control of other countries, it could be restricted, taken or kept from those that need it. The data could also be compromised because of different

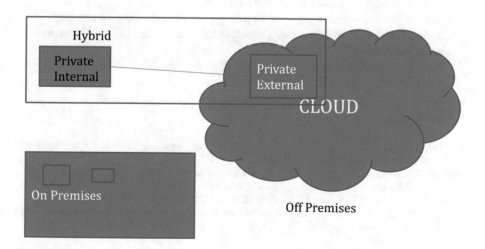

Fig. 11.3 Cloud deployment types. *Source* Wikipedia

legal systems and changes in government policies. We see such problems in multi-national firms quite often.

- *Downtime*: how much can any vendor guarantee ongoing service without outage, and to what extent can Clouds have failover abilities and at what cost?
- *Conversions*: If you should need to move from one Cloud provider to another, what are the risks and problems with moving applications and what compatibility problems will occur especially with data conversion?

11.14.1   Summary

This chapter examined the complexities of whether to make or buy an application solution. Hybrid solutions, which involve both a third-party package and internally developed applications, are very realistic alternatives. The evolution of open source and cloud computing offer attractive opportunities to design and create hybrid systems that provide broader alternatives than historically were available. Furthermore, the concept of Driver/Supporter provides a more scientific way of determining whether to make or buy software solutions all of which are consistent with IoT and blockchain architectures to support mobile environments.

11.15   Problems and Exercises

1. Explain what is meant about Core versus Content.
2. Why is TOC so important?
3. Describe Langer's theory of Driver/Supporter.
4. What is the relationship between Driver/Supporter and Make versus Buy?
5. Define Open Source. How can Open Source change the decision process on Make versus Buy?
6. What is Cloud computer? Explain the different types of Clouds.

References

Bradley, S. P., & Nolan, R. L. (1998). *Sense and respond: Capturing value in the network era.* Boston: Harvard Business School Press.

Burt, D. N., Dobler, D. W., & Starling, S. L. (2003). *World class supply management: The key to supply chain management* (7th ed.). Boston: McGraw-Hill/Irwin.

Inman, R. A., Sale, R. S., Green, K. W., & Whitten, D. (2011). Agile manufacturing: Relation to JIT, operational performance and firm performance. *Journal of Operations Management, 29,* 343–355.

Langer, A. M. (2011). *Information technology and organizational learning: Managing behavioral change through technology and education* (2nd ed.). Boca Raton, FL: Taylor & Francis.

Moore, G. (2002). *Living on the fault line: Managing for shareholder value in any economy.* New York: HarperCollins.

The Analyst and Project Management in the Next Generation

12

12.1 Introduction

This chapter provides guidance on system development life cycle methodologies and best practices for project management of in the next generation of systems. Project organization including roles and responsibilities are covered. There are many aspects of the next generation (5G, IoT, Blockchain) that are generic; however, there are certainly many unique aspects when managing these mobile-based systems. Thus, this chapter provides an understanding of where these unique challenges occur in the life cycle of software development. It also focuses on the ongoing support issues that must be addressed to attain best practices.

A project manager who comes from a traditional software development background and understands the phases of software development will perhaps do fine in overseeing the progress of packaged software projects. Indeed, traditional project managers will focus on budget, the schedule, the resources, and the project plan. Unfortunately, packaged software systems, because of their wide spread involvement with many parts of the business, needs to go beyond just watching and managing the software development process. That is, the project management of IoT requires much more integration with the internal and consumer communities. It must combine traditional development with business creation, and because of the pre-existing nature of the package, it also delves into the internal organizations structure and requires their participation in every phase of the development and implementation cycle. It is for these reasons that I advocate that the traditional analyst considers transitioning their skills to include project management. The reasons for my position relates mostly to the addition of the consumer interface. Most traditional project managers are from the software development side, whereas the new generation of development discussed in this book is more about establishing consumer perspectives. Listed below are some of the unique components of mobile development projects.

© Springer Nature Switzerland AG 2020
A. M. Langer, *Analysis and Design of Next-Generation Software Architectures*,
https://doi.org/10.1007/978-3-030-36899-9_12

1. *Project Managers as Complex Managers*: packaged software projects require multiple interfaces that are outside the traditional user community. They can include interfacing with writers, editors, marketing personnel, customers and consumers, all who might be stakeholders in the success of the system.
2. *Shorter and Dynamic Development Schedules*: Due to the dynamic nature of packaged systems, the development is less linear. Because there is less experience and more stakeholders, there is a tendency to underestimate the time and cost to complete.
3. *New Untested Technologies*: There is so much new technology offered particularly for Web developers that there is a practice of using new versions of development software that has not matured. The method of obtaining new software is easily distributed over the Web, so it is relatively easy to try new versions as soon as they become available. We are also in the world of DevOps which supports the corrections of applications after their release—its fixing and ongoing development on the fly!
4. *Degree of Scope Changes*: Mobile applications, because of their involvement with many aspects of the consumer, tend to be much more prone to scope creep because of the predictive nature of the requirements. Project managers need to work closely with internal users, customers, and consumers to advise them of the impact of changes on the schedule and the cost of the project. Unfortunately, scope changes that are influenced by changes in market trends may not be avoidable. Thus, part of a good strategy is to manage scope changes rather than attempt to stop them—which might not be realistic.
5. *Costing packaged Systems is Difficult*: The software industry has always had difficulties in knowing how to cost a project. Third-party systems are even more difficult because of the number of variables, unknowns, and use of new technologies and procedures. Blockchain products, IoT and cloud will likely be dominated by various outsource and vendor packages.
6. *Lack of Standards*: The software industry continues to be a profession that does not have a governing body. Thus, it is impossible to have real enforced standards as other professions enjoy. While there are suggestions and best practices, many of them are unproven and not kept current with new developments. Because of the lack of successful packaged software projects, there are few success stories to create new and better best practices.
7. *Less Specialized Roles and Responsibilities*: The software development team tends to have staff members that have varying responsibilities. Unlike traditional software projects, separation of roles and responsibilities are more difficult when operating in a mobile environment For example, defining the exact role of a analyst can be very tricky; for example, are analysts programmers, database developers, or content designers? The reality is that all of these roles can be part of a developer's responsibility.
8. *Who Bears the Cost?* There is general uncertainty as to who should bear the cost of the packaged system. This refers to the internal organization of stakeholders who need to agree on the funding. This becomes even more complex when there are delays and cost overruns, because the constituents cannot easily

agree on who is at fault and therefore who should bear the burden of the additional costs.

9. *Project Management Responsibilities are very Broad*: Mobile architectures have broader management responsibilities and need to go beyond those of the traditional IT project manager. Working with third-party interfaces require management services outside the traditional software staff. As discussed in Chap. 1 analysts need to interact more with external users as well as with non-traditional members of the development team such as content managers and social media staff. Therefore, there are many more obstacles that can cause project managers to fail at their jobs.

10. *The Product Never Ends*: The nature of how applications are built today and deployed suggests that they are living systems. This means that they have a long-life cycle made up of ongoing maintenance and enhancements. So, the traditional begin and end project does not apply to a packaged software project that inherently must be implemented in ongoing phases.

Figure 12.1 summarizes these differences between traditional and packaged software projects.

The questions that need to be answered are not limited to what the process and responsibilities should be, but also who should do them? It is my position that a business analyst takes the responsibility of managing the process from inception to completion. The duties and responsibilities of a business analyst are excellent prerequisites for understanding the intricacies of project management. Their roles as analysts require them to have relationships with the organization and an understanding of the politics and culture that drives the business. I am not suggesting that

Packaged Software Projects	Traditional Projects
Project managers are not always trained client managers	Different
Development project schedules tend to be short	Similar
New and untested third-party software are often implemented	Usually never
Changes in scope occur during implementation	Similar
Pricing model does not really exist	Different
Standards for package production do not exist	Similar
Team roles are less specialized	Different
Users have difficulty bearing the costs of development, especially during planning	Different
Project manager responsibilities are broad	Different

Fig. 12.1 Next generation software and traditional projects compared

every analyst should become a project manager, but rather that one of the analysts should also be the project manager. In order to determine the right fit, it is important to define the skill sets that are required for successful project management. These are summarized below:

- Software experience.
- Understanding of budgeting, scheduling, and resource allocation.
- Excellent written and verbal communication skills.
- Ability to hold and lead meeting discussions.
- Detailed oriented yet globally motivated (can see the difference between the forest and the trees).
- Pragmatic.
- A sense of humor that comes across as a natural personal trait as opposed to an acted one.
- Ability to be calm and level-headed during crisis.
- Experience with Web technologies, multimedia, and software engineering.

Unfortunately, it is difficult to find the project manager that has all of these traits. In many cases it is wise to promote from within and develop the expertise internally. This is especially effective because an internal individual is a known quantity, and most likely already fits into the culture of the organization. Most important is that the individual is accepted in the culture. On the downside, it takes time to develop internal talent, and sometimes this trained talent leaves the company once they have received their training. There is benefit to bringing in someone from the outside because they can have a fresh view of the project, and offer more objective input on what needs to be done to get the project finished on time.

12.2   Defining the Project

The first step for the project manager is to develop a mission statement for the project. A mission statement helps managers and users/consumers to focus on three core tasks:

1. Identify the projects objectives.
2. Identify the users and consumers.
3. Determine the scope of the project.

12.3   Identify Objectives

Project objectives are defined as the results that must be attained during the project. According to Lewis (1995), project objectives must be specific, measurable, attainable, realistic, and time-limited. The most difficult of these objectives tends to be "measurable" and "attainable." Ultimately, objectives state the desired outcomes and focus on how the organization will know when it is reached? Objectives are typically devised by the project's stakeholders. These individuals are usually executives and managers that have the most to gain from the successful implementation of the packaged software system. Unfortunately, while this sounds good, it is difficult to implement. In reality, it is difficult for executives to articulate on what they are looking for. Indeed, the packaged software paradigm has simply forced many executives to create products because they think it is a competitive advantage for their companies to have one. This, in essence, means that executives might be driven by the fear that they must do something, or something is better than nothing.

Packaged software objectives evolve and cause many iterative events to occur, especially in the early phases of the project. Good and effective objectives tend to be short sentences that are written down. Using this format, objectives can be used by project managers to effectively avoid scope creep. The objectives should be distributed to all stakeholders and project members so that everyone understands them.

12.4   Identify Users

Chapter 2 covered the significance of users and their importance to the success of any project. In the chapter I defined three types of users: internal, customers and consumers. It is important for project managers to understand the value of the input from each of these users. Indeed, the content of the Web site will ultimately be determined by the users who access the site. However, managers and developers often disagree on how much input is needed from users. This is further complicated when managing packaged software projects because of the diversity of the users and the complexity of decisions that must be made. Furthermore, there is always limited time, so packaged software project managers need to be as productive as possible with how user input is obtained, the types of interviewing that is done, and the method of measuring the value of the user's input.

Since the only way to measure a site's success is to determine whether the objectives have been met, the philosophy of who gets interviewed and how much value their input has should be mapped to the original objectives set forth by management. Thus, besides the internal users, the real obstacle for project managers is to identify which users know best what they want from the packaged software system. Besides one-on-one interviews, the project manager can also obtain information from two other sources: market research and focus groups.

There are many firms that provide market research services. Such firms have databases of researched information relating to user preferences and behaviors. They also collect information about packaged solutions and what users expect from them. Every packaged software system should have a budget using a market research firm so that they can obtain an objective and independent opinion about user preferences, particularly within a certain market segment. Conducting a focus group is a cheaper yet effective way to get objective input from users. It is particularly useful when attempting to assess consumer preferences. Focus groups involve the selection of sample consumers that the project manager feels represents the typical user. The sessions are filmed behind a mirror, and users respond to questions about their preferences when using a packaged software system. The focus group typically needs a moderator who controls the meeting agenda and ensures that all of the research questions are answered by the participants. During all sessions it is important that the project manager ensure that the objectives of the packaged software system be clearly defined to the audience. The objectives should be in writing and reviewed before the start and end of each session. In addition, the objectives should be written on a whiteboard or flip chart so participants can be reminded of the scope of the project should certain users start discussions on tangent subjects.

12.5   Determining the Scope of the Project

The scope of the project relates to the time and budget of when it needs to be completed. Because there is always a limited amount of time and money to create product, the scope of the project must be negotiated against what can be done with what users want done. Thus, scope is the domain of functions and features that will be included in the packaged software system based on a specific time commitment and cost outlay. The best approach to formulating a scope statement is to first create a work breakdown structure that contains the mission statement, lists the objectives, and formulates the tasks and subtasks to complete each objective. Thus, a work breakdown structure is really a form of functional decomposition of the tasks necessary to meet the objectives. Once stakeholders and the project manager agree on the objectives and what tasks will be done to attain them, then the scope of the project is complete. Figure 12.2 depicts a sample work breakdown structure.

Once tasks and subtasks have been determined, the packaged software project manager needs to determine the time and cost of completing each component. Thus, the work breakdown structure will eventually contain the costs for each task within each objective for the entire project as proposed. Management and the project manager can then begin the process of negotiating what can be completed on time and on budget by removing subtasks or tasks as appropriate.

Another valuable approach to building packaged software projects is phasing deliverables. Because packaged software projects tend never to be finished, it might be advisable to deliver some portion of the system first and then add-on functionality in subsequent releases of the system. Obviously, this might not always be feasible; there are third-party software systems that cannot be phased, that is, they are all or nothing at all. However, I believe that all projects can have some level of phased development, and that such development in the long run benefits the entire scope of the project. Indeed, first releases of a packaged solution typically need

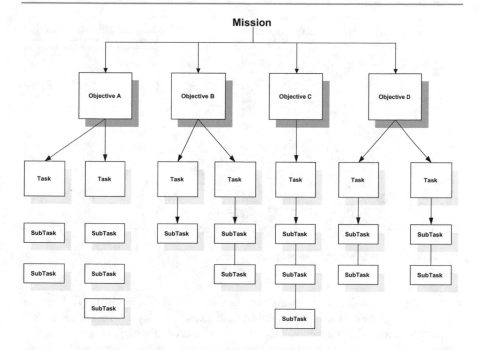

Fig. 12.2 Sample work breakdown structure

revision anyway, and the second phase or version might be a better time to add certain features and functions.

I stated that the final work breakdown schedule represents the scope of the project. Typically, the packaged software project manager will finalize the scope statement by preparing a document that includes the work breakdown structure and articulates how this structure will be formulated into deliverables for the project. In many ways the scope document acts as a management report and reiterates the mission and objectives of the project along with the project plan. Figure 12.3 shows a typical project plan developed in Microsoft Project.

12.6 Managing Scope

A project plan can sometimes be referred to as a work breakdown schedule or WBS. As shown in Fig. 12.3, it depicts every step in the project and can enforce dependencies within tasks and subtasks. This is important, because changes to the plan may affect other tasks. A WBS product like Microsoft's Project provides an automated way of tracking changes and determining its effect on the entire project. A popular method of tracking changes is called "critical path analysis." Critical path analysis involves the monitoring of tasks that can have an effect on the entire scope

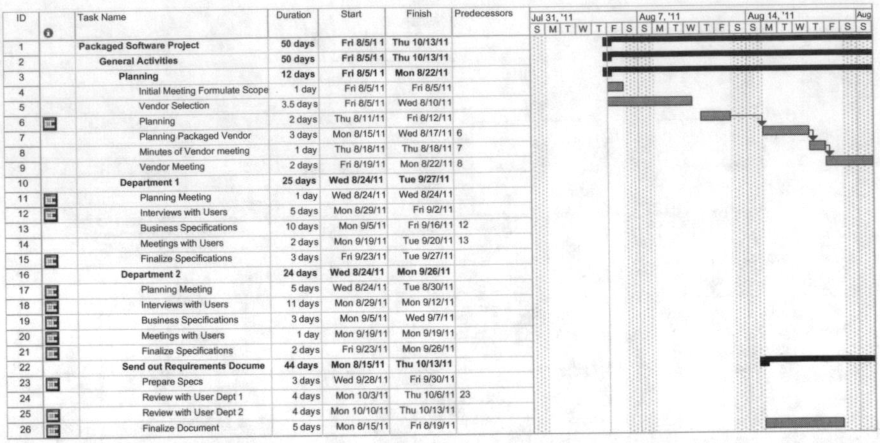

ID		Task Name	Duration	Start	Finish	Predecessors
1		**Packaged Software Project**	50 days	Fri 8/5/11	Thu 10/13/11	
2		**General Activities**	50 days	Fri 8/5/11	Thu 10/13/11	
3		**Planning**	12 days	Fri 8/5/11	Mon 8/22/11	
4		Initial Meeting Formulate Scope	1 day	Fri 8/5/11	Fri 8/5/11	
5		Vendor Selection	3.5 days	Fri 8/5/11	Wed 8/10/11	
6		Planning	2 days	Thu 8/11/11	Fri 8/12/11	
7		Planning Packaged Vendor	3 days	Mon 8/15/11	Wed 8/17/11	6
8		Minutes of Vendor meeting	1 day	Thu 8/18/11	Thu 8/18/11	7
9		Vendor Meeting	2 days	Fri 8/19/11	Mon 8/22/11	8
10		**Department 1**	25 days	Wed 8/24/11	Tue 9/27/11	
11		Planning Meeting	1 day	Wed 8/24/11	Wed 8/24/11	
12		Interviews with Users	5 days	Mon 8/29/11	Fri 9/2/11	
13		Business Specifications	10 days	Mon 9/5/11	Fri 9/16/11	12
14		Meetings with Users	2 days	Mon 9/19/11	Tue 9/20/11	13
15		Finalize Specifications	3 days	Fri 9/23/11	Tue 9/27/11	
16		**Department 2**	24 days	Wed 8/24/11	Mon 9/26/11	
17		Planning Meeting	5 days	Wed 8/24/11	Tue 8/30/11	
18		Interviews with Users	11 days	Mon 8/29/11	Mon 9/12/11	
19		Business Specifications	3 days	Mon 9/5/11	Wed 9/7/11	
20		Meetings with Users	1 day	Mon 9/19/11	Mon 9/19/11	
21		Finalize Specifications	2 days	Fri 9/23/11	Mon 9/26/11	
22		**Send out Requirements Docume**	44 days	Mon 8/15/11	Thu 10/13/11	
23		Prepare Specs	3 days	Wed 9/28/11	Fri 9/30/11	
24		Review with User Dept 1	4 days	Mon 10/3/11	Thu 10/6/11	23
25		Review with User Dept 2	4 days	Mon 10/10/11	Thu 10/13/11	
26		Finalize Document	5 days	Mon 8/15/11	Fri 8/19/11	

Fig. 12.3 Third-party software project plan

of the project, meaning it can change the timeframe and cost of delivery. A critical path is defined as a task that if delayed will cause a delay in the entire project. A task that can delay the project is then called a critical task. The importance of managing critical tasks is crucial for successful management of packaged software projects. Project managers are often faced with the reality that some task has slipped behind schedule. When faced with this dilemma. The project manager needs to decide whether dedicating more resources to the task might get it back on schedule. However, the first thing that the project manager needs to assess is whether the task can affect the critical path. If the answer is yes, then the project manager must attempt to use other resources to avoid a scope delay. If the task is not critical, then the delay may be acceptable without needing to change the project plan. Figure 12.4 shows a critical task and a non-critical task.

12.7 The Budget

Budgeting is one of the most important responsibilities of a project manager. The budget effectively prices the tasks that must be delivered and rolls them up to the project cost level. It is important to recognize that all budgets are estimates. Therefore, they are never 100% accurate—if they were, they would not be budgets. The concept behind budgeting is that some tasks will be over-budget and others will be under-budget resulting in an offset that essentially balances out to the assumptions outlined in the original plan. Budgets are typically built on expense categories. Figure 12.5 shows the common budget categories that projects should be tracking.

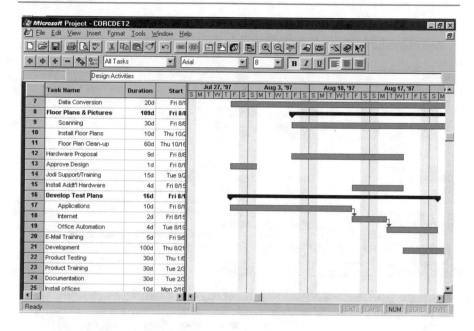

Fig. 12.4 Critical and non-critical tasks using Microsoft project

As stated above, a project budget is a set of assumptions. Typical budget assumptions are:

- All content will be provided in machine-readable form.
- The content manager will approve content design within 24-hours.
- The Web design team will present two alternate design schemas.
- Graphics for the Web site are finalized and ready for integration.

It is not a bad idea for the project manager to create a budget document that includes a list of the assumptions because it allows the manager to track whether incorrect assumptions caused delays in the scope of the project. Unfortunately, there are also hidden costs that tend not to be included in project budgets. The following is a list of common hidden costs that are missed by the project manager:

- Meetings.
- Phone calls.
- Research.
- Development of documents and status reports.
- Project administration.
- Review Sessions.
- Presentations to management.

Project Budget Sample   12/12/2010

Description		Low	High	Comments				
Hardware								
Servers	$	150,000	200,000					
Workstations		400,000	600,000					
Modems		25,000	30,000					
Scanners		5,000	20,000					
Total Hardware		580,000	850,000					
Software								
Base Product		250,000	350,000	Cost of vendor base product software				
Database		100,000	250,000	Database vendor software license				
Office Automation		40,000	100,000	Forms elimination and e-mail intranet				
CAD/CAM-Scanning		40,000	100,000	Scanning software				
Media Production		90,000	150,000					
				Modification costs to vendors product to meet				
Software Modifications		75,000	125,000	needs				
Total Software		595,000	1,075,000					
Services								
Network and Software Design		5,000	8,500	Analysis & design of system				
				Specialists that may be				
Consultants		240,000	285,000	needed				
Scanning Documents		125,000	135,000	Service to put all plans in system				
Conversion of Data		120,000	130,000					
Installation		45,000	87,000	low: 300x150 high 300x250 plus servers				
Training		25,000	85,000	Train the trainer all train everyone				
Total Services		560,000	730,500					
Total	$	1,735,000	2,655,500					

Fig. 12.5 Sample E-business project budget expense categories

As previously discussed, some project managers add a 10–15% cushion to their budgets to absorb common hidden costs. While I do not support cushions, it is acceptable if actually listed as a budget item, as opposed to a cushion on each budget line item.

12.8   The Project Team

The project team is unique from other traditional project organizations. Most of the significant differences are attributable to the addition of the packaged software responsibilities. Today, the roles and responsibilities for developing and supporting packaged software team are far more complex and specialized. The project team has evolved because business managers understand the importance of technology in transforming the way business is done. On the other hand, there are certainly traditional roles and responsibilities that have not changed and are generalizable across any software development project.

While the structure of project teams can vary depending on the type of project, the size of the system, and the time to complete, typical organizations contain the following roles and responsibilities:

- *Project Manager*: The project manager is responsible for the scope of work, developing the project plan, scheduling, allocating resources, budgeting, managing the team, interfacing with users, and reporting to management on progress. The project manager also deals with politics and other business issues, which include but are not limited to contract negotiations, licensing of third-party products, and staff hiring. In some instances, the project manager is responsible for handling customer and consumer needs as it relates to the design and development of the packaged software system. Perhaps the most important responsibility of the packaged software project manager is to know at all times what has been done and what needs to be done.
- *Account Manager*: The account manager is usually a senior manager who is responsible for a number of projects. Account managers also serve clients in a number of ways, from selling new product to providing client support. In many ways, the account manager is the representative of the client's needs to the internal development team. Account managers can be called upon to obtain information from customers about their needs and their feedback on how the system supports their needs.
- *Technical manager*: This individual is the senior technologist of the project. He/she is usually from the development team and is the most experienced developer. The technical manager is responsible for ensuring that the correct technology is being used and deployed properly. This individual manages the programmers, database developers and other system integrators. The technical manger provides feedback on the development status of each task and reports to the project manager.
- *Programmer*: A programmer may be needed to do certain custom modifications if not outsourced. He/she is responsible for coding applications for the project. These applications are coded to spec and can include a myriad of technologies including but not limited to server-scripts, database applications, applets and ActiveX controls. Development languages used on third-party software projects vary, but the Web uses such languages as Java, JavaScript, Visual Basic, VBScript, SQL, and C/C++. The technical manager usually manages this individual; however, large projects may employ multiple levels of programmers. In certain situations, junior programmers report to senior developers who act as mentors to them.
- *Business Analyst*: This individual is responsible for gathering all of the user requirements and designing the logic models and architecture of the system, which include process models, data models, transactions system design, and

process specifications. Ultimately the business analyst is responsible for site architecture, navigation, search and data retrieval, and interaction design. This role is sometimes called an Information Architect.

- *Designer*: Designers create the look and feel of the screens. They use various tools to design template content and overall screen structure. Screen designers report to the project manager who sets the overall project philosophy.
- *Database Administrator* (*DBA*): The DBA is responsible for all physical database design and development. This individual must also fine-tune the database to ensure efficient operation. Other responsibilities include data partitioning, data warehouse setup, data replication and report generation.
- *Network Engineers*: These individuals are responsible for designing network configurations that support the packaged software system. Sometimes a network engineer is also a security specialist who is responsible for registering domain names, setting up email servers and chat rooms.
- *Security Expert*: While this might be the network specialist, larger packaged software projects employ a dedicated security expert who works with encryption formulas, integrates with content systems, and focuses on securing online transactions. This individual can also advise the project manager on strategies to implement certain component applications.
- *Quality Assurance Specialist*: This individual is responsible for creating test scripts to ensure that the packaged software system operates within spec. The purpose of the test plans, sometimes called acceptance test plans, are to provide the minimal set of tests that must be passed for the site to go into production. The reporting structure for quality assurance personnel varies. Some believe that it is part of the development team and therefore should report to the technical manager while others believe that it needs to report directly to the project manager. There are still others that believe that the quality assurance staff should report to the chief information officer (CIO). The reporting structure is dependent on how highly the department values the independence of the testing function.
- *Tester*: A tester is simply a person who carries out the test plans developed by the quality assurance staff. These people typically report to quality assurance staff, however, sometimes testers are users who are working with the account manager to assist in the testing of the product. These testers are in a sense a beta test user (beta test means that the software is tested in a live environment).

While this is an exhaustive list of potential positions to have on a project, it is unlikely that every one of these positions will exist in any given third-part software project. In reality, it is more important that the functionality of each of these positions is carried out, regardless of who takes on the responsibility. In order to provide better insight to the roles and responsibilities of each position, Fig. 12.6 reflects a comparison matrix that compares roles with responsibilities.

Figure 12.7 identifies the necessary skills of each of the critical members of the project team.

Type of Project Activity	Staff Organization
Marketing and Media	Account executive, project manager, creative manager, designer, copywriter, production artist, quality assurance specialist
Project transactions	Project manager, technical manager, analyst, creative manager, database administrator,, tester
Input/Output data	Project manager, technical manager, creative manager, database administrator, analyst, Web production specialist, quality assurance specialist
Mobile	Project manager, technical manager, creative manager, network engineer, internet production specialist, quality assurance specialist

Fig. 12.6 Project team roles and responsibilities matrix

Skill Set	Description
Project Management	Ability to communicate with staff, executives, and users. Keep project on schedule.
Architecture	Ability to design the user interface and interact with technical development personnel
Graphic Design	Ability to transform requirements into visual solutions
Graphic Production	Ability to develop efficient graphics for Web browser use
Content Development	Ability to design and develop text and interactive content
Programming	Ability to use HTML, Python, Javascript, etc.

Fig. 12.7 Necessary project skill sets

12.9 Project Team Dynamics

It is not unusual for third-party software development projects to operate from multiple sites with multiple interfaces. This may require much more organization and communication among the members of the team.

12.10   Set Rules and Guidelines for Communication

When project members are separated by locations or even by work schedules it is very important that everyone know what their roles are, and what everybody else's responsibilities are to each other. Project managers should establish guidelines on communication and require each member to provide short statuses on where they are in the project. While this might seem like a bother to team members, I have found that it provides immense value to the project manager because it forces each person to discipline themselves and report on where they are in their respective worlds. Furthermore, it forces project staff to articulate in writing what they have accomplished, what is outstanding, and what they plan to get done. Figure 12.8 shows a sample status report that tracks previous objectives with current objectives. It should not take more than 15 min to complete.

12.11   Review Sites

Project managers should create extranets that allow project staff to view the work of the project team. This also allows work to be approved virtually over the Web. Furthermore, status reports and general announcements can be viewed by authorized members of the project team, as well as by stakeholders and users. Often extranet documents can be coupled with regular conference call sessions where project members can openly discuss reports, Web design samples, and determine new milestones as appropriate. Unfortunately, managing an extranet review site requires overhead and someone who can do the work. This simply means that someone on the project's staff or some assigned administrative person needs to take on the responsibility. There is also the challenge of dealing with staff that do not comply with procedures, or need to be reminded about delivering their status reports on a regular basis. While this is unfortunate, it is a reality. However, it also tells the project manager who needs to be watched closer than other team members.

12.12   Working with User Resources

Users are an interesting yet challenging resource. While they are clearly needed to perform reviews and quality assurance, they are not officially assigned to the project and are therefore not under the control of the project manager. Unfortunately, this can cause problems if a user is not responsive to the needs of the project team. This can be very damaging because the staff will be dependent on receiving timely feedback from these resources. Lack of responsiveness from a user can also alienate them from the project team. Indeed, there is nothing more damaging than a user that shows disinterest in the project that is being designed to serve them. Therefore, the packaged software project manager needs to be careful on what commitments are

XYZ CORP
Project Plan

Project Status Report

Date 1/31/11	Vendor: XSP	Consultant: Art Langer

Previous Objectives: Objective	Previous Target Date	Status
Send out Software Proposals	1/23/19	Done
Send out Hardware Bids and check labor rates	1/25/19	2/6/19
Get XSP to modify detail design contract to include specifics on deliverables.	1/23/19	Done, waiting approval
Speak with product consultant and contract with him for data and layouts	1/25/19	2/3/19
Finalize Network Design Requirements Document	1/23/19	2/3/19

New Project Objectives: Objective	Target Date
Get specification on hardware certification machines from XSP	2/05/19
Get room to do interviews and reviews	2/10/19
Ensure that accounting features are integrated in specs	2/10/19
Art to interview Joe	2/03/19
Check ability of MAS to create output files for display on Web	2/06/19
Identify and meet with brokers as part of interviews	2/03/19
Get prices for Scanning of floor plans for Web browsing	2/10/19

Attendees	A. Langer & Assoc.	Art Langer	XSP

Fig. 12.8 Sample project status report

made when a user resource is made available to them. Obviously, the project manager would want the individual to report directly to him/her and be a full-time resource. This could also be wishful thinking. It certainly can be dangerous to turn down the help. While this may appear to be a catch-22, it can be managed. First, the packaged software project manager should establish the need for user assistance early in the project and include it in the requirements documentation (it should be part of the assumptions section). Second, the project manager may need to limit the amount of work assigned to the user if they indeed are just part-time resources.

12.13 Outsourcing

As discussed in earlier chapters, not all packaged projects can be completed using internal staff. In fact, most are not. Using outsourced resources makes sense for many projects, especially those that might need very unique and qualified personnel that is not on staff, or not deemed worthy of full-time employment. There may also be a lack of talent, which is the usual reason why consultants are hired in the first place. Sometimes, outsourced relationships are managed as strategic partnerships, which means that an outside business provides specific services for the firm on a regular basis. Project strategic partnerships can be made in many different areas or phases of the project. Outsource firms can provide video or audio engineers as needed, or network support personnel to assist in installing the packaged solution. On the other hand, the entire project should not be outsourced because there may be a false sense of comfort that the firm's personnel need not be involved in the responsibility. I believe in the long run this is a mistake. Remember that outsourced firms have their own destiny and growth to manage.

12.14 Planning and Process Development

In order to operationalize the project plan and to meet the budget, it is necessary to develop a phased implementation guide that helps members of the team understand where they are in the process. Unfortunately, the project task plan is much too detailed to use, so it is a good idea for the project manager to develop a higher-level document that can be used during the development project. Such a plan typically contains four phases: (1) Strategy, (2) Design, (3) Development, and (4) Testing. Figure 12.9 reflects the major software phases and includes the activities and output of each phase.

1. *Strategy*: This phase requires that the stakeholders, account executive, users, and the project manager all meet to agree on the objectives, requirements, key milestones, and needs of the target audience. The activities in this phase comprise of the goals and objectives, feedback from users, research from outside sources, and the project proposal document. The culmination of these steps should be summarized in a document to be used by the project team. This document is sometimes referred to as a Creative Brief. The creative brief is really a summary of the original proposal in template form and created so that project team members can quickly refer to it and obtain the information they need. Figure 12.10 depicts a Creative Brief Template.

   Creative briefs allow the project manager to conduct effective brainstorming sessions with staff. The document acts as an agenda, and also a checkpoint to ensure that discussions are not going beyond the objectives of the project and that the products are designed in accordance with the target audience's needs.

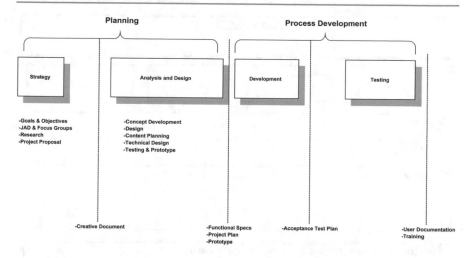

Fig. 12.9 Packaged software development phases

2. *Design*: Design represents the second major phase. It is comprised of all the tasks that participate in the design of the user interface, the analysis and design of technical specifications, and the overall architecture of the packaged software system. The results of the design phase are the functional specifications, detailed project plan timeline and budget, and Web site and report prototypes. The design phase typically requires that stakeholders and other users to review and sign-off on the specification document before the project can proceed. Of course, the approval process of the design document can be done in phases, that is, portions of the document can be approved so that the implementation can go forward on some limited basis. During the process of designing the packaged system, it is important that the project team have access to the content of all screens and programs. This is especially important for large projects where there can be more than one design group creating content. As stated in the previous chapter, content development is an iterative process; therefore, the team members need to constantly have access to the current site architecture and schematic designs. That is why it is so important to have the technical specifications in a CASE tool and content system so that all members can have instant access to the current state of the project.

3. *Development*: This phase includes all of the activities that are involved with actually building the site. The challenge for project managers is to control changes made to the original specifications. This typically occurs after the first prototype reviews where users begin to change or enhance their original requirements. While it is not impossible to change specifications, it is certainly dangerous and can be a major cause of scope creep.

4. *Testing*: During this phase developers and users are testing the site and reporting errors. Errors are tricky issues; they must be classified in particular areas and

Creative Brief Template		
XYZ, INC		

Dates:	Dept:	User:

Target Audiences:

Project Scope

Objectives:

Image: (Explain the image that the system must convey to users)

Current Brand: (Explain the current brand and image of the business to its users)

User Experience: (Explain the most important thing that users should experience from using the system)

\gminutes Copyright © 1995-2019
A. Langer & Assoc., Inc

Fig. 12.10 Sample creative brief

levels of severity. For example, some errors cause an application to abort, which would be considered a critical error. Others might be aesthetic in nature and can be scheduled for fixing but are not severe enough to holdup going live. Still other errors are not really errors, but rather deficiencies in design. This means that the program is performing to the specification but not in the way the user really expects. All of these issues need to be part of an overall test plan, which identifies what types of errors are critical and how they affect the development process. Figure 12.11 shows a typical test plan.

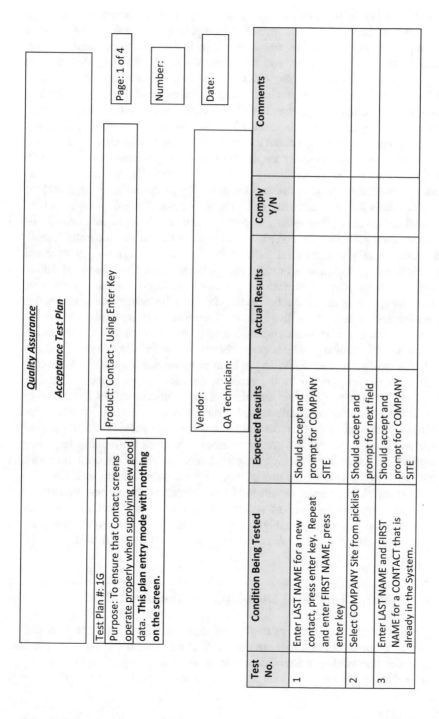

Quality Assurance

Acceptance Test Plan

Test Plan #: 1G

Purpose: To ensure that Contact screens operate properly when supplying new good data. **This plan entry mode with nothing on the screen.**

Product: Contact - Using Enter Key

Page: 1 of 4

Number:

Vendor:

QA Technician:

Date:

Test No.	Condition Being Tested	Expected Results	Actual Results	Comply Y/N	Comments
1	Enter LAST NAME for a new contact, press enter key. Repeat and enter FIRST NAME, press enter key	Should accept and prompt for COMPANY SITE			
2	Select COMPANY Site from picklist	Should accept and prompt for next field			
3	Enter LAST NAME and FIRST NAME for a CONTACT that is already in the System.	Should accept and prompt for COMPANY SITE			

Fig. 12.11 QA test plan

12.15   Technical Planning

Technical planning is the process where the project team develops a working strategy for how they will use the features of the packaged system. These features include all of the components of development including database, programming, transaction systems, multimedia, and scripting. Technical planning is simple in concept; how do all of the technology pieces come together, how do they interface, and what is the schedule of implementation? Because of the object-oriented methods that are employed by packaged software systems, it is easy to have components developed in separate teams of programmers. However, there comes a time when all components must come together and interface with each other. When the components interface correctly the system works. Project managers never quite know if interfaces will work until they are actually tested. The concept of "working" means many things in software development. The obvious definition is that the program performs its tasks correctly and to specification. There is another part of what "working" means. This relates to performance of the application. While an application might calculate the correct output, it may not do so efficiently. Application performance problems tend to first show up during component interface testing. Fortunately, this is at a time where applications can be fine-tuned before they go into production. Unfortunately, many interface-oriented performance problems first appear in the production system because the testing environment was not a true representation of the live system. Therefore, it is important for project managers to ensure that the test system correctly matches the live environment. Indeed, many performance problems occur because of the unexpected stress load on the system. Notwithstanding when an application performance problem is discovered, the main challenge is to fix the problem. Sometimes performance interface problems can be a serious problem especially if the solution requires a redesign of the application architecture or substantial changes to the network infrastructure. Any serious problems of this caliber could cause serious setbacks to the project schedule and its cost. In many ways the planning and decision making regarding how applications are designed, which program languages should be used, and what network platforms to choose are crucial steps for the project team to make. The project manager must attempt to surround him-/herself with the best knowledge available. This knowledge base of people might not exist in the organization. Therefore, the project manager might need to seek guidance from third-party consultants who can act as specialists during these critical decision times.

12.16   Defining Technical Development Requirements

A large part of whether systems are implemented properly has to do with how well the detailed technical requirements are prepared. The technical project team needs to define the technical requirements to implement the logical specifications. Remember that logical specifications do not necessarily specify what hardware or software to use. Thus, the technical team must evaluate the logical specification and make recommendations on how the actual technical specifications are to be built. The project manager can be more effective if they ask some key questions:

- *Are we thinking of using technologies that we have not used before?* Using unknown technologies can be very dangerous. An unknown technology is not only a new product; it is a product that the development team has never used. Because of the extent of new developments in technology, dealing with unknown hardware and software needs to be addressed and risks assessed by the project manager.
- *What benefits will be derived from new technologies?* Implementing new technologies for the sake of new technologies is not a good reason to implement unknowns. This concept gets back to the old cliché: "if it's not broken don't fix it!"
- *What type of coding is being done?* This relates to whether program code is being developed from scratch or via modified software packages. Each has its advantages and disadvantages. Writing code from scratch takes longer, but provides the architecture that best fits the design. Package software is faster to develop, but may not fit well with the overall needs of the company. The rule of thumb is never to modify packages by more than 20% of their total code. When this percent is exceeded the benefit realized form the package is so minimum that developers might as well start from scratch.
- *Will there be access to production-like testing environments?* This was covered earlier. Project managers must ensure that the proper testing facilities are available to mirror the production environment.

12.17 Maintenance

Packaged software projects should never be implemented without considering how to preserve maintainability. Maintainability is a universal concept that relates to what defines a quality product. Products that work is one thing, those that work and are maintainable is another. In packaged software systems, product that cannot be maintained easily is problematic. I have previously discussed the power of content management systems and CASE software as vehicles to support maintenance of packaged software systems. There are other best practices that need to be performed during the planning stage. First, the manner in which code will be developed and the standards to be upheld needs to be agreed upon and put in writing. Technical managers should also define how they intend to enforce these standards. Documentation of code should also be addressed in the documents. Furthermore, there needs to be agreement on the database design as well. This involves getting the Database administrators to agree on limits to de-normalization, naming standards, and the methods of coding stored procedures and database triggers.

Another important component of maintenance quality is planning for growth. The issue of growth relates more to network infrastructure than software development. First, the project manager needs to address issues of hardware scalability. This relates to the capacity of the network before the hardware architecture needs to be changed to accommodate new applications. Second, database servers must be configured with real-time backup architecture (no single point of failure concepts) and data warehouses need to be designed to perform at peak times.

12.18   Project Management and Communication

Successful project managers communicate well, not only with their staffs, but with vendors, management, and users. Indeed, sometimes communications skills are more important than technical ones. Obviously, a complete project manager has both. However, the advent of packaged software systems has placed even more emphasis on the importance of communication within the project team. There are many reasons why poor communications occurs during project life cycles. According to Burdman (1999), there are 11 leading causes for communication problems on a project team.

1. *People come from different disciplines*: Communication is difficult enough among those that work together every day. The influx of many different disciplines on an packaged software project creates more challenges because staffs are not as familiar with each other. Remember that relationships are very important for team interaction. Many teams need to spend time just getting acclimated to each other's business styles.
2. *Lack of mutual understanding of the technology*: Project members do not have consistent understandings of the technology. For example, some staff might use the word "table" to define a logical database entity, while others call it a file. The best solution to this problem is to distribute a list of common technology definitions to all project participants.
3. *Personalities*: This occurs in all projects. Some people have conflicting personalities and do not naturally get along.
4. *Hidden agenda*: There are often political agendas that team members have. They are sometimes difficult to assess, but they definitely cause problems with communication among project staff. These individuals are set in their ways and have questionable dedication to the success of the project, that is, they have a more important political agenda.
5. *Ineffective meetings*: Meetings for the sake of meetings is no reason to meet. Sometimes too many meetings can be counterproductive to getting things done. It can sometimes be a false solution to other problems that exist in the project team. Some meetings are necessary but run too long, and participants begin to lose focus on the agenda. Project managers need to be cognizant of the time allotments they make to meetings and to respect those timeframes.

6. *Proximity*: The demographics of where project team members reside is obviously a factor in hindering communications among the team members. While this is a disadvantage, teleconferencing and video conferencing are all possible anecdotes for managing communication projects at a distance. Communication can be further hindered because of long distances between staff especially when there are time zone changes. In these situations, even conferencing is not feasible. Usually the best way of communicating is through email and extranets.

7. *Assumptions*: Team members can often make assumptions about things that can cause communication breakdowns. Usually assumptions create problems because things that are believed to be true are not written down.

8. *Poor infrastructure and support*: The severity of this problem is often overlooked. It includes the frustrations of having computer troubles, email incompatibilities, and other hardware failures that contribute to communications problems. The best approach to avoiding these frustrations is for the project manager to be very aggressive on having them fixed timely and properly.

9. *Being an expert*: Every project has one or two "know-it-allers" who attempt to dominate meetings and want to orchestrate their point of view on the rest of the staff. These individuals spend so much time telling others what to do that they forget what they have to do to make the project successful. Project managers should be very aggressive with these type of participants by making it clear what everyone's role is, including the project manager's responsibilities!

10. *Fear*: Fear is a very large barrier with certain staff. These staff members become overwhelmed with the size, complexity, and length of packaged software project and it can cause them to lose their perspective and creativity. Project managers need to interject and provide assistance to those members who struggle with a packaged software system.

11. *Lack of good communications structure*: Good communications systems fit–in with the culture of the organization and are realistic in what they might accomplish. Many communications problems exist simply because the infrastructure does not relate to the needs of the staff.

12.19   Summary

This chapter attempted to provide a project manager with a perspective on the salient issues that can help them be successful. This chapter was not intended to provide a complete step-by-step approach to managing complex projects. I included this chapter because I believe that many analysts can also serve as excellent project managers. Indeed, much of the important issues discussed in this chapter relate to many of the skills that analysts must have to perform their responsibilities as software engineers. These include:

- *Communications Skills*: Analysts have significant experience in working with users to obtain input so they can develop system requirements properly.

- *Meeting Management*: JAD sessions are more complex meetings than typical project meetings. Analysts that have also been JAD facilitators are very well trained on how to control meetings.
- *Politically Astute*: Analysts are experienced with working with people who have hidden agendas and are driven by politics.
- *Technically Proficient*: Analysts are educated in logic modeling and are familiar with much of the technical issues that come up during the project life cycle.
- *Project Planning*: Analysts are used to developing project plans and managing to deliverables; each analysis and design task can be seen as a mini-project.
- *Documentation*: Analysts are supporters of good documentation and understand the value of having maintainable processes.
- *Executive Presence*: Analysts work with executive users and understand how to interact with them.
- *Quality Assurance*: Analysts are familiar with quality assurance test plans and testing methodologies. They are often involved with test plan development.

12.20   Problems and Exercises

1. Describe five unique aspects of packaged software projects.
2. Compare packaged products with traditional ones.
3. What is a Mission Statement for a project mean? How should a project manager define the project mission?
4. Explain how the scope of a project is determined and how it is controlled.
5. What are the key categories of a project budget?
6. What are the roles and responsibilities of the project team? Who determines the members?
7. Explain the components of a project status report? How often should the report be issued and to whom?
8. What is a creative brief? What are the key components?
9. What are the important aspects of project management communication?
10. Why are packaged software decisions so important for IoT and blockchain applications?

References

Burdman, J. (1999). *Collaborative web development*. New York: Addison-Wesley.
Lewis, M. (1995). *Project management: 25 popular project management methodologies*. New York: Amazon Digital Services LLC.

Conclusions and the Road Forward

13

The purpose of this book was to provide a roadmap for the reader on the processes and considerations for building new computer architectures to support the emerging consumer market. Many of us believe that the advent of 5G will be a major technical development that will dramatically accelerate the way businesses and individuals use and depend on digital technologies. The book has integrated two essential challenges to designing and developing these new systems: the technical components that comprise the physical machines and devices; and the various approaches to migrating the vast number of legacy systems that run the day-to-day organizations throughout the world.

The legacy challenge seems overwhelming and I believe that many organizations will fail to understand the urgency of why they need to move quickly to digital. They likely will fail to embrace 5G and IoT timely and as a result I expect further business failures in the future. Indeed, it seems every few months another business giant announces performance shortfalls. I mentioned in Chap. 6 the GE Digital division underperformance. The results of this failure have devastated GE's market cap and for the first time they have fallen from the Dow 500, the last of the original companies that started on the Index! There are others, most recently Toy-R-Us, which seemed to have a good private equity partner but could not get its organization to embrace a digital culture fast enough. Although expectations were positive, they inevitably closed. I also expect many other businesses to consider merging especially in the higher education marketplace where we have already seen 20 college closures as online education begins to make an impact on what was once thought of as an untouchable business. Another critical dilemma is the reality of the required investment that is needed to transform to digital. I expect the costs to be substantial and I predict that many companies may need to raise cash from private equity partners to appropriately move into the digital age.

Perhaps the most significant message is the need to evolve an organization's culture. It's one thing to have the technology infrastructure, it's another to have the right people. We see many companies struggling to find the appropriate technical talent, and even more cannot get their existing staffs to embrace digital. Prahalad

© Springer Nature Switzerland AG 2020
A. M. Langer, *Analysis and Design of Next-Generation Software Architectures*,
https://doi.org/10.1007/978-3-030-36899-9_13

and Krishnan (2008) foresaw this development over 10 years ago when they created the "New House of Innovation" shown in Fig. 13.1.

This model understood that the consumerization of technology established the
requirements for organizations to develop "flexible and resilient business processes
and focused analytics" (p. 6). The key of this need, according to the authors, are
based on two key "pillars" defined as N = 1 and R = G. It means that every
business must serve each consumer as a unique individual (N = 1). R = G, states
that resources must be global in order to accomplish N = 1. So the House of
Innovation defines consumerization of IT based on an organization's ability to
satisfy a customer's needs by providing global service offerings. Therefore, to
successfully compete in the digital world businesses must provide agile services
24/7 365 to their consumers! So, if a consumer needs service at off-peak times,
organizations will likely need to provide support using global resources to survive.
For example, a consumer needing support at midnight in New York might by
serviced by someone in India—it's an "on-demand," model that can be compared
to the Burger King slogan, "have it your way!" The point is that consumers want it
on demand. So, Burger King would have to provide hamburgers at midnight and
offer an unlimited number of choices. Not an easy challenge, but this is the reality
of what consumerization of technology means. My message is the need to treat
every consumer as a unique person and to deliver services seamlessly. To
accomplish this objective, the analyst needs to provide software to support agility—
the type of agility that is now available because of 5G speed, IoT devices, and
blockchain architecture. This book has also advocated the expansion of the analyst's role to help manage the cultural shift necessary to become a digital company.

Fig. 13.1 Prahalad and Krishnan's New House of Innovation

Indeed, a digital entity needs to assimilate new types of employees that can integrate with the legacy staff. Snow et al. (2017) posited that digital organizations were necessary to support the new technologies and that such organizations needed to be designed for self-authorization rather than the traditional hierarchical methods that project control and coordination. They concluded that successful organizations need workplaces that parallel digital uncertainties and recommend structures that are "highly engaged and productive" (p. 11).

The analyst can help achieve this agile and digital culture by ensuring that applications cater to the following four fundamentals:

1. Speed is more important than cost.
2. The organization must be empowered to respond to consumer and market needs (N = 1).
3. Applications must work on all devices while conforming to standardization or internal controls.
4. The digital architecture must be designed to maximize consumer options and on-demand responsiveness to users.

In addition, the new digital paradigm discussed in this book must deal the following realities:

- The expansion of smartphones, social networks, and other consumer technologies are creating the need to change cultures, attitudes, and workplace practices.
- Features to deal with vulnerable technology and information—particularly security and reliability.
- Increasing pressure for quality and efficiency—while keeping costs low.
- Rise of data-driven decision making for critical systems.
- New approaches to innovation—rethinking how to provide and control new products and services.
- Consideration for disruptive disasters that are caused by man-made catastrophes and wars.

13.1 Sense and Response and the End of Planning

The reality of living in the present, not the future, is simply uncertainty. How can we architect applications and networks that can scale and deal with this uncertainty? It was Accenture in 2012 who first published the report, "Reimagining Enterprise IT for an Uncertain Future" that emphasized the inability to do traditional planning because of the rise of data-driven decision-making for critical systems design. The report also called for new approaches to innovation; the rethinking of how to design new products and services that could be delivered to the consumer at lower costs. Bradley and Nolan brilliantly published their book in 1998 "Sense and Respond" and predicted the shifts in consumer behavior that became a reality

20 years later. They posited that organizations needed to sense opportunities and respond with strategies as opposed attempting to predict the future using budgets.

Today we see the theory of quantum physics being mapped to computing systems. The hope is that new architectures will be able to utilize resources differently in each instance as opposed to the way a binary system does sequential calculations. This means that applications may be able to do complex calculations incredibly faster. This will provide the backbone systems to support a whole new transition of human behavior—behavior that changes how we function in a digital-driven world. So, what do we do with the notion of budgets, and how far forward can we really budget the future? In other parts of my work I have emphasized that all budgets need to be adjustable during the year, and multi-year budgets appear to be an exercise of false hopes. The larger challenge is to somehow marry human behavior with systems architecture. I discussed this idea in the book referencing, The Inversion Factor (2017), which suggested the importance of the shift from product to function. Cusumano et al. (2019) present the concept of a business as a platform. They define platforms as bringing "together individuals and organizations so they can innovate or interact in ways not otherwise possible, with the potential for nonlinear increases in utility and value" (p. 13). Simply put, the better platform will beat the best product.

13.2   The Role of Artificial Intelligence and Machine Learning

As stated in the book, The AI Advantage (Davenport 2018), "in the short run AI will provide evolutionary benefits; in the long run, it is likely to be revolutionary (p. 7). Ultimately, I believe that applications will be easier to use and will support better decision making. ML, on the other hand, will make the larger impact over the next decade. In 2017 Deloitte surveyed 250 people that they deemed competent cognitive managers and found that 58% of them were already utilizing ML in their business practices. I have no doubt, as I have advocated in this book, that IoT will serve to accelerate higher percentages of ML use globally. The reasons are quite simple, the physical mind can no longer handle the size of datasets that will be accumulated by collection devices. Humans will have to rely more on intelligent machines to give them results.

Historically, aggressive adopters of new technologies are tech-based vendors who see opportunities early in the marketplace. As such, the traditional firms are laggards as it takes them more time to get over the risk factors, a sort of watch and wait syndrome, and then begin to think about ways to transform their enterprises. Such is the same with digital disruption, as we see a slow adoption of AI and ML in most organizations. Unlike historical revolutions, this concerns me because of the acceleration of change produced by new technologies. Being a laggard during this period could ensure failure. The reader should recall the concept of a shrinking s-curve that requires organizations to act quicker if they are to compete in a global economy that is being swamped by new startups every day.

However, there are real advantages that large companies have if they embrace AI and ML, because they have lots of historical data and may have the financial reserves to support a firm-wide integration. Most companies according to Davenport (2018) see AI affecting three fundamental business activities:

1. Automating repetitive work processes via automation.
2. Obtaining better perspectives and understanding by examining structured data using ML.
3. Getting a better understanding of customer and employee behavior via chatbots and ML.

Ultimately the entire AI and ML phenomena will require companies to become more "cognitive" and accepting that the "inversion factor" must be key to their survival, or as Gupta (2018) states "define your business around your customers, not your products or competitors" (p. 17). The analyst must be a key operator in this effort. Although there is no question that c-level executives must initiative the effort and formulate the strategies, the management and implementation of these efforts are far more critical to overall adoption. Success cannot be attained without having individuals that have a core understanding of computer architecture, and project management.

13.3   Blockchain

As I explained in the book, blockchain is at the center to implementing the next generation of technology. 5G is the initiator, and IoT the aggregator from a physical perspective. But nothing really can operate safely without a more security focused processor. While blockchain is still in its infancy, it remains our best option. All of the criticisms are real, especially as it relates to latency. I predict a plethora of new blockchain products in the market, likely all to be very specialized within a particular industry. Without doubt we will see new blockchain products offered by third-party vendors that will integrate legacy systems similar to the way we combined object and relational databases. Further, vendors will develop applications that will link their cloud products with existing legacy applications and offer software migration products. I expect this integration and migration to be at the forefront for CIOs over the next 15 years. Most concerning is the future of the mainframe, which never seems to meet its s-curve end. The good news is that it remains one of the most secure systems, the bad is that there are diminishing staff available. Mainframes are also very expensive to operate and require large facilities. The bottom line is that companies who rely on mainframe processing need to consider an alternative hardware platform if they are to remain competitive.

13.4 Cloud and More Cloud

While blockchain appears to be the new processing engine, cloud is the future data repository. The two must interact in a very complex way. Indeed, blockchain architecture wants to replicate data and store it, and cloud wants to somehow still maintain central control. On the other hand, cloud also represents some combination of the two since we can execute applications from the cloud in addition to data storage. So, what cloud really does is to provide a more cost-effective way to expand applications across the internet. I have covered many cloud models and choices, all of which have advantages and disadvantages, but inevitably cloud will mature into a major hosting facility that allows global businesses to thrive throughout the world, assuming that of course that security can be assured.

13.5 Quantum Computing

I wanted quantum computing to be part of this book because I believe it has the potential to succeed. Many of us in the industry recognize that silicon needs to be replaced if we are to take advantage of the sophisticated applications that are available but too slow to implement. That is, software has outpaced the evolution of hardware. This means that further advancements will be limited if we cannot increase processing speed. Unfortunately, quantum is not yet available or feasible. We will have to see what the next generation of hardware architecture provides to support more blockchain products and sophisticated hashing algorithms to secure an abundance of data produced by IoT and used for AI and ML.

13.6 The Human Factor of the Next Generation of the Digital Organization

I have published and presented about how organizations will operate in the digital age. We have many names that define this challenge—how to be agile, are you a digital-born company or a west-coast culture. Over the next 5 years many companies will be faced with the dilemma of how to integrate executives from the baby-boomer era; with Gen-X line—managers who supervise the day-to-day operations; and with millennials or Gen Y operating staff.

The current prediction is that 76 million Baby Boomers (born 1946–1964) and Gen X workers (born 1965–1984) will be retiring over the next 10 years. The question for many corporate talent executives is how to manage this transition.

Baby Boomers still inhabit most executive positions in the world. The average age of CEOs is 56 and 65% of all corporate leaders are Baby Boomers. Organizations over the next 5 years will need to produce career paths that will be attractive to Millennials. Therefore, Baby Boomers and Gen X's need to:

- acknowledge some of their preconceived perceptions of current work ethics that are simply not relevant in today's complex environments.
- allow millennials to be promoted to satisfy their ambitions and temper their sense of entitlement.
- be more open to more flexible work hours, offer telecommuting, and develop a stronger focus on social responsibility.
- support more advanced uses of digital capabilities at work, including those used in the millennial personal world.
- train more senior staff to help millennials to better understand why there are organizational constraints at work.
- provide more professional reviews and feedback.
- implement programs that improve the verbal communications skills of millennials who seem more comfortable with nonverbal text-based methods of communication.
- implement more continual learning and rotational programs that support a vertical growth path for younger employees.
- explain the legacy systems and give millennials an opportunity to participate in transforming strategies.

In summary, it is up to the Baby Boomer and Gen X leaders to evolve their styles of management to attract millennials in traditional companies. The difficulty of attracting new talent in these companies has become a major challenge in the last five years. The challenge to accomplish this objective is not trivial given the wide variances on how these three generations think, plan, take risks and most important learn.

13.7   Transforming to a Digital Enterprise

Zogby completed an interactive poll of 4,811 people on perceptions of different generations. Forty-two percent of the respondents stated that the Baby Boomers legacy would be remembered for their focus on consumerism and self-indulgence. Gen Y, on the other hand, are reported to be more self-interested, entitled narcissists who want to spend all their time posting "selfies" to Facebook. Other facts offer another perception of these two generations as shown in Table 13.1.

Table 13.1 Baby Boomers versus millennials. *Source* Langer (2018)

Baby Boomers	Gen Y
Married later and less children	Not as aligned to political parties
Spend lavishly	More civically engaged
More active and selfless	Socially active
Fought against social injustice, supported civil rights, and defied the Vietnam War	Cheerfully optimistic
Had more Higher Education access	More concerned with quality of life than material gain

Table 13.2 Management
roles 2008–2013

Baby Boomer (ages 49–67)	19%
Gen X (ages 33–48)	38%
Gen Y (18–32)	87%

Research completed by Ernst and Young (2014) offers additional comparisons among the three generations as follows:

1. Gen Y individuals are moving into management positions faster due to retirements, lack of corporate succession planning, and their natural ability to use technology at work. Table 13.2 shows percentage comparisons between 2008 and 2013.
   The acceleration of growth to management positions among Gen Yers can be further illuminated in Table 13.3 by comparing the prior five-year period from 2003 to 2007.
2. While responders of the survey felt Gen X were better equipped to manage than Gen Y, the number of Gen Y managers is expected to double by 2020 due to continued retirements. Another interesting result of the research related to millennial expectations from their employers when they become managers. Specifically, millennial managers expect: (1) an opportunity to have a mentor; (2) receive sponsorship, (3) have more career-related experiences, and (4) receive training to build their professional skills.
3. Seventy-five percent of respondents that identified themselves as managers agree that managing the multiple generations are a significant challenge. This was attributed to different work expectations and the lack of comfort with younger employees managing older employees.

Table 13.4 provides additional differences among the three generations.

The message in this section is that the architectural transformation from 5-G to blockchain and cloud, while essential is not enough without the assimilation of new cultures and should not be underestimated as a critical issue in any corporate digital strategy.

Table 13.3 Management
roles 2003–2007

Baby Boomer (ages 49–67)	23%
Gen X (ages 33–48)	30%
Gen Y (18–32)	12%

Table 13.4 Baby Boomer, Gen X and millennial compared. *Source* Langer (2018)

Baby Boomers	Gen X	Millennials (Gen Y)
Seek employment in large established companies that provide dependable employment	Established companies no longer a guarantee for lifetime employment. Many jobs begin to go offshore	Seek multiple experiences with heavy emphasis on social good and global experiences. Re-evaluation of offshoring strategies
Process of promotion is well defined, hierarchical and structured eventually leading to promotion and higher earnings—concept of waiting your turn	Process of promotion still hierarchical, but based more on skills and individual accomplishments. Master's degree now preferred for many promotions	Less patience with hierarchical promotion policies. More reliance on predictive analytics as the basis for decision-making
Undergraduate degree preferred but not mandatory	Undergraduate degree required for most professional job opportunities	More focus on specific skills. Multiple strategies developed on how to meet shortages of talent. Higher Education expensive and concerns increase about the value of graduate knowledge and abilities
Plan career preferably with one company and retire. Acceptance of a gradual process of growth that was slow to change. Successful employees assimilated into existing organizational structures by following the rules	Employees begin to change jobs more often given growth in the technology industry and opportunities to increase compensation and accelerate promotion by switching jobs	Emergence of a "GIG" economy and the rise of multiple employment relationships
Entrepreneurism was seen as an external option for those individuals desiring wealth and independence and willing to take risks	Corporate executives compensation dramatically increases no longer requiring starting businesses as the basis for wealth	Entrepreneurism promoted in Higher Education as the basis for economic growth given the loss of jobs in the US

13.8 Security is a Central Problem

While there are a number of initiatives to increase monitoring of potential threats to security, there needs to be more efforts to integrate quality assessments of how systems need to be redesigned or substantially improved. Just as the traditional analyst engaged in the design of acceptance test plans. In essence design aspirations and performance requirements must be balanced against security exposure. In this book I have articulated the historical error of not considering enough of the security exposure as we leaped forward with the internet. Looking back, it seems so obvious of what was needed. Now we must take a step back and rethink how we balance all of the new capabilities of IoT and big data with the protection of the consumer. I do not believe that GDPR is the ultimate solution. Below are some of the evolving challenges that analysts need to consider:

- Learning how to articulate the need to insert security during design with senior and operating managers. Ultimately business leaders need to recognize that security may indeed limit wanted features and functions.
- How to work with internal users to roll out new and scalable security-minded cultures.
- Assisting in developing procedures for dealing with a compromised application and replacement with alternative processes.

Let's examine these three issues further. When speaking with senior managers, analysts need to relate to business objectives and avoid too much technical jargon. Analysts also need to explain how third-party vendors will be managed to ensure that there are proper controls over security exposure. Analysts should also apply benchmarks of what competitors are doing to convince their managers to adhere to best security design practices. Peer pressure is often an effective practice to push change.

In reference to security cultures, it's important to explain the risks associated with poor internal staff practices. Indeed, many compromises have been linked to careless work behaviors of staff. Analysts should consider discussing additional security steps in the SDLC. Such steps should include documentation of how security considerations have been addressed during analysis and design. Analysts need; however, to ensure that the security implications are weighed against its effects on business operations and competitive advantage.

13.9   The Role of the Analyst

This book has presented several aspects of how digital disruption is driving the need for a new generation of computer architecture. The chapters integrated the technical design ramifications of building these new architectures and ways to transition existing legacy systems. I have also advocated that the digital era will require major changes in organization design including new roles and responsibilities. Most important I have defined the expanding role for those professionals that will be engaged in the actual analysis, design, and management of these new systems. Figure 13.2 provides a diagram of the complex functions that need to be performed by managers and analysts.

Figure 13.2 suggests that it is unlikely that one individual could take on all of these duties. I believe these new responsibilities will result in the creation of a new organization focused on all of these aspects. I have referred to these staff members as "analysts" throughout the book. Whether this is their inevitable title, I believe the existing business and systems analysts are the likely individuals to take on these roles and to build the new organization. Ultimately the name of the position is less

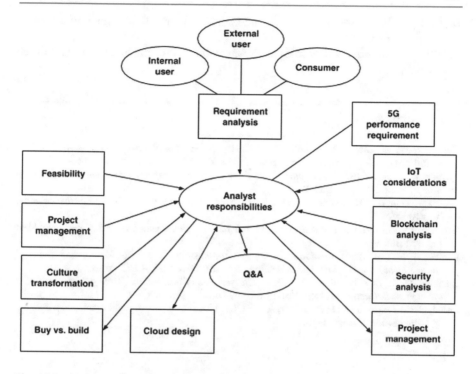

Fig. 13.2 Analyst responsibilities for mobile-based systems

important than having the staff that can carry out the necessary requirements to build and support these new mobile systems. I think it is essential that organizations give high priority to these functions in order to successfully transition their firms to meet the demands of tomorrow's consumers.

13.10 Problems and Exercises

1. Explain the New House of Innovation and its relevance in supporting consumer-based mobile systems.
2. What is meant by the "social architecture" of a firm and its relationship to organizational culture?
3. What is meant by N = 1 and R = G?
4. How does "sense and response" challenge the notion of planning and prediction?
5. What is the concept of a digital organization? What are the behaviors of such organizations?

6. Compare Baby Boomers, Gen X, and Gen Y staff and discuss the importance of their synergy.
7. What are the crucial new roles and responsibilities for the analyst of the future? Explain.

References

Cusumano, M. A., Gawer, A., & Yoffie, D. B. (2019). *Business of platforms: Strategy in the age of digital competition, innovation, and power*. New York: Harper Collins Publishers.

Davenport, T. H. (2018). *The AI advantage: How to put the artificial intelligence revolution to work*. Cambridge, MA: MIT Press.

Ernst and Young. (2014). *Gen Y managers perceived as entitled, need polish*. https://www.cnbc.com/100997634.

Gupta, S. (2018). *Driving digital strategy: A guide to reimagining your business*. Cambridge, MA: Harvard Business Review Press.

Langer, A. M. (2018). *Information technology & organizational learning: Managing behavioral change through technology and education* (3rd ed.). New York: CRC Press.

Prahalad, C. K., & Krishnan, M. S. (2008). *The new age of innovation: Driving cocreated value through global networks*. New York: McGraw-Hill.

Snow, C. C., Fjeldstad O. D., & Langer, A. M. (2017). Designing the digital organization. *Journal of Organization Design*, 1–13.

Glossary

5G The fifth generation cellular network technology. The industry association 3GPP defines any system using "5G NR" (5G New Radio) software as "5G", a definition that came into general use by late 2019 (Wikipedia)

Acceptance Test Plans Set of tests that if passed will establish that the software can be used in production

Actuator A component of a machine that is responsible for moving and controlling a mechanism or system, for example by opening a valve. In simple terms, it is a "mover" (Wikipedia)

Alternate Key An attribute that uniquely identifies a row or occurrence in an entity. An alternate key cannot be the primary key

Application Program Interface (API) An interface or communication protocol between a client and a server intended to simplify the building of client-side software. It has been described as a "contract" between the client and the server, such that if the client makes a request in a specific format, it will always get a response in a specific format or initiate a defined action (Wikipedia)

Artificial Intelligence (AI) The intelligence demonstrated by machines, in contrast to the natural intelligence displayed by humans (Wikipedia)

Attribute A component of an entity or object. An attribute may or may not be an elementary data element

Blockchain Is a growing list of records called *blocks*, that are linked using cryptology. Each block contains a cryptographic hash of the previous block, a timestamp and transaction data. (Wikipedia)

Business Process Re-Engineering (BPR) A method to redesign existing applications

Business Specification A document which reflects the overall requirements of a process or system written in a prose format. The focus of the business specification is to provide the user with enough information so they can authorize the development of the technical requirements

© Springer Nature Switzerland AG 2020
A. M. Langer, *Analysis and Design of Next-Generation Software Architectures*,
https://doi.org/10.1007/978-3-030-36899-9

CASE (Computer Aided Software Engineering) Products which are used to automate and implement modeling tools and data repositories

Case A format for developing application logic in a process specification

Client An application that request services from applications

Cloud computing An on-demand availability of computer system resources, especially data storage and computing power, without direct active management by the user. The term is generally used to describe data centers available to many users over the Internet. Large clouds, predominant today, often have functions distributed over multiple locations from central servers. If the connection to the user is relatively close, it may be designated an edge server (Wikipedia)

Crow's Foot A method of showing the relationship or association between two entities

CRUD Diagram An association matrix that matches the types of data access between entities and processes. CRUD represents Create, Read, Update, and Delete

Cyber Security The protection of computer systems from the theft of or damage to their hardware, software, or electronic data, as well as from the disruption or misdirection of the services they provide. The field is becoming more important due to increased reliance on computer systems, the Internet and wireless network standards such as Bluetooth and Wi-Fi, and due to the growth of "smart" devices, including smartphones, televisions, and the various devices that constitute the "Internet of things"

Data Dictionary (DD) A dictionary that defines data. A component of the data repository

Data Flow Component of a data flow diagram that represents data entering or leaving a process, external or data store

Data Flow Diagram (DFD) A tool that shows how data enters and leaves a process. A data flow diagram has four possible components: data flow, data store, external, and process

Data Repository A robust data dictionary that contains information relating to data element behavior

Data Store Component of a data flow diagram that represents data that can be accessed from a particular area or file. A data store is sometimes called "data-at-rest."

Data Warehousing A de-normalized database created to focus on decision support activities. Data warehouse hold historical information and cannot be used to update data

Elementary Data Element A functionally decomposed data element

Entity An object of interest about which data can be collected. Entities can consume a number of attributes

Entity Relational Diagram (ERD) A diagram that depicts the relationships among the stored data

Equal Rights The rights can be set to be equal among all minors of the chain

External Component of a data flow diagram which represents a provider or user of data that is not part of the system. Externals are therefore boundaries of the system

Facilitator An impartial individual responsible for the controlling the flow of JAD sessions

Functional Decomposition The process for finding the most basic parts of a system

Functional Overview Subset view of a specification. The subset usually covers a particular function of the system

Functional Primitive A functionally decomposed data flow diagram

Gantt Chart Tool that depicts progress of tasks against time. The Chart was developed by Henry L. Gantt in 1917

Immutability The events of an object in a blockchain cannot be changed, so that an audit trail of transactions is traceable

Internet of Things (IoT) Is a system of interrelated computing devices, mechanical and digital machines, objects, animals or people that are provided with unique identifiers (UIDs) and the ability to transfer data over a network without requiring human-to-human or human-to-computer interaction (Wikipedia)

ISO 9000 International Organization for Standardization, quality standard 9000

Job Description Matrix The portion of an individual's job description that strictly focuses on the procedural and process aspects of the individual's position

Key An attribute of an entity or database that uniquely identifies a row, occurrence or record

Key Business Rules Business rules of key attributes that are enforced at the database level (as opposed to the application level)

Legacy System An existing automated system

Leveling Functional decomposition of a data flow diagram. Each decomposition is called a "level."

Logical Data Modeling (LDM) A set of procedures that examines an entity to ensure that its component attributes should reside in that entity, rather than being stored in another or new entity

Logical Equivalent An abstraction of the translation from physical requirements to software

Long Division An abstraction of the relationship of arithmetic formulas to functional decomposition

Machine Learning (ML) The scientific study of algorithms and statistical models that computer systems use to perform a specific task without using explicit instructions, relying on patterns and inference instead (Wikipedia)

Metadata Data about the data being sent or received in a client/server network

Non-repudiation The identity of the author of a transaction are guaranteed among all members of the blockchain

Normalization The elimination of redundancies from an entity

Open Systems Standards in applications software that allow such software to run across multiple operating system environments

Pre–Post Conditions A format for developing application logic in a process specification

Predictive Analytics Encompasses a variety of statistical techniques from data mining, predictive modelling, and machine learning, that analyze current and historical facts to make predictions about future or otherwise unknown events (Wikipedia)

Primary Key A key attribute that will be used to identify connections to a particular entity. Normalization requires that every entity contain a primary key. Primary keys can be formed by the concatenation of many attributes

Process A function in a data flow diagram in which data is transformed from one form to another

Process Specification A document that contains all of the algorithms and information necessary to develop the logic of a process. Process specifications can be comprised of the business and programming requirement documents. Process specifications are sometimes called "minispecs."

Program or Technical Specification A technical algorithm of the requirements of a process or system

Prototype A sample of a system that does not actually fully operate. Most software prototypes are visual depictions of screens and reports. Prototypes can vary in capability, with some prototypes having limited functional interfaces

Pseudocode A generic or structured English representation of how real programming code must execute. Pseudocode is a method used in the development of process specifications

Quantum Computing The study of a still-hypothetical model of computation. Whereas traditional models of computing such as the Turing machine or Lambda calculus rely on "classical" representations of computational memory, a quantum computation could transform the memory into a quantum superposition of possible classical states. A quantum computer is a device that could perform such computation (Wikipedia)

Reverse Engineering The process of analyzing existing applications and database code to create higher-level representations of the code

Robust Software that operates intuitively and can handle unexpected events

Sensor A device, module, machine, or subsystem whose purpose is to detect events or changes in its environment and send the information to other electronics, frequently a computer processor. A sensor is always used with other electronics (Wikipedia)

Server An application that provides information to a requesting application

Spiral Life Cycle Life cycle that focuses on the development of cohesive objects and classes. The spiral life cycle reflects a much larger allocation of time spent on design than the waterfall approach

State Transition Diagram (STD) A modeling tool that depicts time dependent and event driven behavior

Transparency All members or minors of the blockchain are aware of changes

Waterfall System Development Life Cycle A life cycle that is based on phased dependent steps to complete the implementation of a system. Each step is dependent on the completion of the previous step

Bibliography

Adapting to the data explosion: Ensuring justice for all. In *Proceedings of the 2009 IEEE International Conference on Systems, Man, and Cybernetics*, 2009.

Aldrich, H. (2001). *Organizations evolving*. London: Sage.

Allen, F., & Percival, J. (2000). Financial strategies and venture capital. In G. S. Day & P. J. Schoemaker (Eds.), *Wharton on managing emerging technologies* (pp. 289–306). New York: Wiley.

Allen, F., & Percival, J. (2003). Financial strategies and venture capital. In L. M. Applegate, R. D. Austin, & F. W. McFarlan (Eds.), *Corporate information strategy and management* (2nd ed.). New York: McGraw-Hill.

Allen, T. J., & Morton, M. S. (1994). *Information technology and the corporation*. New York: Oxford University Press.

Applegate, L. M., Austin, R. D., & McFarlan, F. W. (2003). *Corporate information strategy and management* (2nd ed.). New York: McGraw-Hill.

Applegate, L. M., McFarlan, F. W., & McKenney, J. L. (1999). *Corporate information systems management: The challenges of managing in an information age*. New York: McGraw-Hill.

Argyris, C. (1993). *Knowledge for action: A guide to overcoming barriers to organizational change*. San Francisco: Jossey-Bass.

Argyris, C., & Schön, D. A. (1996). *Organizational learning II*. Reading, MA: Addison-Wesley.

Argyris, C., Putnam, R., & Smith, D. (1985). *Action science*. San Francisco: Jossey-Bass.

Arnett, R. C. (1992). *Dialogue education: Conversation about ideas and between persons*. Carbondale, IL: Southern Illinois University Press.

Bal, S. N. (2013). Mobile web–enterprise application advantages. *International Journal of Computer Science and Mobile Computing, 2*(2), 36–40.

Batten, J. D. (2002). *Tough-minded management* (3rd ed.). Eugene, OR: Resource Publications.

Bazarova, N. N., & Walther, J. B. (2009a). Virtual groups: (Mis)attribution of blame in distributed work. In P. Lutgen-Sandvik & B. Davenport Sypher (Eds.), *Destructive organizational communication: Processes, consequences, and constructive ways of organizing* (pp. 252–266). New York: Routledge.

Bazarova, N. N., & Walther, J. B. (2009b). Attribution of blame in virtual groups. In P. Lutgen-Sandvik & B. Davenport-Sypher (Eds.), *The destructive side of organizational communication: Processes, consequences, and constructive ways of organizing* (pp. 252–266). Mahwah, NJ: Routledge/LEA.

Beinhocker, E. D., & Kaplan, S. (2002). Tired of strategic planning? *McKinsey Quarterly, 2*, 48–57.

Bensaou, M., & Earl, M. J. (1998). The right mind-set for managing information technology. In J. E. Garten (Ed.), *World view: Global strategies for the new economy* (pp. 109–125). Cambridge, MA: Harvard University Press.

Benson, J. K. (1975). The interorganizational network as a political economy. *Administrative Science Quarterly, 20*, 229–249.

© Springer Nature Switzerland AG 2020
A. M. Langer, *Analysis and Design of Next-Generation Software Architectures*,
https://doi.org/10.1007/978-3-030-36899-9

Berman, K., & Knight, J. (2008). *Finance intelligence for IT professionals*. Boston: Harvard Business Press.

Bertels, T., & Savage, C. M. (1998). Tough questions on knowledge management. In G. V. Krogh, J. Roos, & D. Kleine (Eds.), *Knowing in firms: Understanding managing and measuring knowledge* (pp. 7–25). London: Sage.

Blackstaff, M. (1999). *Finance for technology decision makers: A practical handbook for buyers, sellers and managers*. New York: Springer.

Boland, R. J., Tenkasi, R. V., & Te'eni, D. (1994). Designing information technology to support distributed cognition. *Organization Science, 5*, 456–475.

Bolman, L. G., & Deal, T. E. (1997). *Reframing organizations: Artistry, choice, and leadership* (2nd ed.). San Francisco: Jossey-Bass.

Bolman, L., & Deal, T. (2003). *Reframing organizations: Artistry, choice, and leadership* (3rd ed.). San Francisco: Jossey-Bass.

Brown, J. S., & Duguid, P. (1991). Organizational learning and communities of practice. *Organization Science, 2*, 40–57.

Brynjolfsson, E., & McAfee, A. (2011). *Race against the machine*. Lexington: Digital Frontier Press.

Brynjolfsson, E., & McAfee, A. (2012). Big data: The management revolution. *Harvard Business Review, 90*(10), 60–68.

Burke, W. W. (2002). *Organization change: Theory and practice*. London: Sage Publications.

Burke, W. W. (1982). *Organization development: Principles and practices*. Boston: Little Brown.

Burns, C. (2009) Automated talent management. *Information Management*. http://www.information-management.com/news/technology_development_talent_management-10016009-1.html.

Bysinger, B., & Knight, K. (1997). *Investing in information technology: A decision-making guide for business and technical managers*. New York: Wiley.

Carr, N. (2003). IT doesn't matter. *Harvard Business Review, 81*(5), 41–49.

Carr, N. G. (2005). *Does it matter? Information technology and the corrosion of competitive advantage*. Cambridge: Harvard Business School.

Cash, J. I., & Pearlson, K. E. (2004, October 18). The future CIO. *Information Week*. http://www.informationweek.com/story/showArticle.jhtml?articleID=49901186.

Cassidy, A. (1998). *A practical guide to information strategic planning*. Boca Raton, FL: St. Lucie Press.

Charan, R. (2006). *Sharpening your business acumen strategy & business*. New York: Booz & Co.

Chesbrough, H. (2003). *Open innovation: The new imperative for creating and profiting from technology*. Cambridge: Harvard Business School.

Chesbrough, H. (2006). *Open business models: How to thrive in the new innovation landscape*. Cambridge: Harvard Business School.

Chesbrough, H. (2011). San Francisco: Jossey-Bass.

Cillers, P. (2005). Knowing complex systems. In K. A. Richardson (Ed.), *Managing organizational complexity: Philosophy, theory, and application* (pp. 7–19). Greenwich, CT: Information Age.

Cohen, A. R., & Bradford, D. L. (2005). *Influence without authority* (2nd ed.). Hoboken, NJ: Wiley.

Cole, R. E. (1985). The macropolitics of organizational change. *Administrative Science Quarterly, 30*, 560–585.

Collis, D. J. (1994). Research note—How valuable are organizational capabilities? *Strategic Management Journal, 15*, 143–152.

Conger, J. (2003). Exerting influence without authority. In L. Keller Johnson (Eds.), *Harvard business update*. Boston: Harvard Business Press.

Cortada, J. W. (1997). *Best practices in information technology: How corporations get the most value from exploiting their digital investments*. Paramus: Prentice Hall.

Croon Fors, A. & Stolterman, E. (2004). Information technology and the good life. In Kaplan, T. et al. (Eds.), *Information systems research. Relevant theory and informed practice.*

Cross, T., & Thomas, R. J. (2009). *Driving results through social networks. How top organizations leverage networks for performance and growth.* San Francisco: Jossey-Bass.

Cyert, R. M., & March, J. G. (1963). *The behavioral theory of the firm.* Englewood Cliffs, NJ: Prentice-Hall.

Deluca, J. (1999). *Political savvy: Systematic approaches to leadership behind-the-scenes.* Berwyn, PA: EBG.

Dewey, J. (1933). *How we think.* Boston: Health.

Dodgson, M. (1993). Organizational learning: A review of some literatures. *Organizational Studies, 14*(3), 375–394.

Dragoon, A. (2002). This changes everything. Retrieved December 15, 2003, from http://www.darwinmag.com.

Earl, M. J. (1996a). *Information management: The organizational dimension.* New York: Oxford University Press.

Earl, M. J. (1996b). Business process engineering: A phenomenon of organizational dimension. In M. J. Earl (Ed.), *Information management: The organizational dimension* (pp. 53–76). New York: Oxford University Press.

Earl, M. J., Sampler, J. L., & Short, J. E. (1995). Strategies for business process reengineering: Evidence from field studies. *Journal of Management Information Systems, 12,* 31–56.

Easterby-Smith, M., Araujo, L., & Burgoyne, J. (1999). *Organizational learning and the learning organization: Developments in theory and practice.* London: Sage.

Edwards, C., Ward, J., & Bytheway, A. (1996). *The essence of information systems* (2nd ed.). Upper Saddle River: Prentice Hall.

Eichinger, R. W., & Lombardo M. M. Education competencies: Dealing with ambiguity. Microsoft in Education|Training. Microsoft. Web.

Eisenhardt, K. M., & Bourgeois, L. J. (1988). Politics of strategic decision making in high-velocity environments: Toward a midrange theory. *Academy of Management Journal, 31*(4), 737–770.

Fahey, L., & Randall R. M. (1998a). Integrating strategy and scenarios. In L. Fahey & R. M. Randall (Eds.), *Learning from the future,* ch. 2. New York: Wiley.

Fahey, L., & Randall R. M. (1998b). What is scenario learning? In L. Fahey & R. M. Randall (Eds.), *Learning from the future,* ch. 1. New York: Wiley.

Ferrell, O. C., & Gardiner, G. (1991). In *Pursuit of ethics.* USA: Smith Collins.

Fineman, S. (1996). Emotion and subtexts in corporate greening. *Organization Studies, 17,* 479–500.

Fisher, D., Rooke, D., & Torbert, B. (1993). *Personal and organizational transformations through action inquiry.* Boston: Edge/Work Press.

Fleming, C., & von Halle, B. (1989). *Handbook of relational database design.* Menlo Park, CA: Addison-Wesley.

Foster, R. N., & Kaplan, S. (2001). *Creative destruction: Why companies that are built to last underperform the market—And how to successfully transform them.* New York: Currency.

Friedman, T. L. (2007). *The world is flat.* New York: Picador/Farrar, Straus and Giroux.

Friedman, T. L., & Mandelbaum, M. (2012). *That used to be us.* London: Picador.

Gardner, C. (2000). *The valuation of information technology.* New York: Wiley.

Garvin, D. A. (1993). Building a learning organization. *Harvard Business Review, 71*(4), 78–84.

Garvin, D. A. (2000). *Learning in action: A guide to putting the learning organization to work.* Boston: Harvard Business School Press.

Gavitte, G., & Rivikin J. W. (2005, April). How strategists really think: Tapping the power of analogy. *Harvard Business Review,* 54–63.

Gephardt, M. A., & Marsick, V. J. (2003). Introduction to special issue on action research: Building the capacity for learning and change. *Human Resource Planning, 26,* 2.

Glasmeier, A. (1997). *The Japanese small business sector* (Final report to the Tissot Economic Foundation, Le Locle, Switzerland, Working Paper 16). Austin: Graduate Program of Community and Regional Planning, University of Texas at Austin.

Goonatilake, S., & Teleaven, P. (1995). *Intelligent systems for finance and business*. New York: Wiley.

Govindarajan, V., & Trimble, C. (2004). Strategic innovation and the science of learning. *MIT Sloan Management Review, 45*(2), 67–75.

Grant, D., Keenoy, T., & Oswick, C. (1996). *Discourse and organization*. London: Sage Publications.

Grant, D., Keenoy, T., & Oswick, C. (Eds.). (1998). *Discourse and organization*. London: Sage.

Grant, R. M. (1996). Prospering in a dynamically-competitive environment—Organizational capability as knowledge integration. *Organization Science, 7*(4), 375–387.

Gregoire, J. (2002, March 1). The state of the CIO 2002: The CIO title, What's it really mean? *CIO*. http://www.cio.com/article/30904/The_State_of_the_CIO_2002_The_CIO_Title_What_s_It_Really_Mean_.

Habermas, J. (1998). *The inclusion of the other: Studies in political theory*. Cambridge, MA: MIT Press.

Halifax, J. (1999). Learning as initiation: Not-knowing, bearing witness, and healing. In S. Glazier (Ed.), *The heart of learning: Spirituality in education* (pp. 173–181). New York: Penguin Putnam.

Hardy, C., Lawrence, T. B., & Philips, N. (1998). Talk and action: Conversations and narrative in interorganizational collaboration. In D. Grant, T. Keenoy, & C. Oswick (Eds.), *Discourse and organization* (pp. 65–83). London: Sage.

Heath, D. H. (1968). *Growing up in college: Liberal education and maturity*. San Francisco: Jossey-Bass.

Hoffman, A. (2008, May 19). The social media gender gap. *Business Week*. http://www.businessweek.com/technology/content/may2008/tc20080516_580743.htm.

Hogbin, G., & Thomas, D. (1994). *Investing in information technology: Managing the decision-making process*. New York: McGraw-Hill.

Huber, G. P. (1991). Organizational learning: The contributing processes and the literature. *Organization Science, 2*, 99–115.

Hullfish, H. G., & Smith, P. G. (1978). *Reflective thinking: The method of education*. Westport, CT: Greenwood Press.

Huysman, M. (1999). Balancing biases: A critical review of the literature on organizational learning. In M. Easterby-Smith, J. Burgoyne, & L. Araujo (Eds.), *Organizational learning and the learning organization* (pp. 59–74). London: Sage.

IBM & Said School of Business, Oxford University. (2012). Analytics: The real-world use of big data in financial services. Retrieved 30 September 2015, http://www-935.ibm.com/services/multimedia/Analytics_The_real_world_use_of_big_data_in_Financial_services_Mai_2013.pdf.

IEEE Access Journal. (2019). https://ieeeaccess.ieee.org/special-sections-closed/modelling-analysis-design-5g-ultra-dense-networks/.

Illbury, C., & Sunter C. (2001). *The mind of a fox* (pp. 36–43). Human & Rousseau/Tafelberg: Cape Town, SA.

Johansen, R., Saveri, A., & Schmid, G. (1995). Forces for organizational change: 21st century organizations: Reconciling control and empowerment. *Institute for the Future, 6*(1), 1–9.

Jones, M. (1975). Organizational learning: Collective mind and cognitivist metaphor? *Accounting Management and Information Technology, 5*(1), 61–77.

Kanevsky, V., & Housel, T. (1998). The learning-knowledge-value cycle. In G. V. Krogh, J. Roos, & D. Kleine (Eds.), *Knowing in firms: Understanding, managing and measuring knowledge* (pp. 240–252). London: Sage.

Kaplan, R. S., & Norton, D. P. (2001). *The strategy-focused organization*. Cambridge, MA: Harvard University Press.

Kegan, R. (1994). *In over our heads: The mental demands of modern life*. Cambridge, MA: Harvard University Press.

Kegan, R. (1998, October). *Adult development and transformative learning*. Lecture presented at the Workplace Learning Institute, Teachers College, New York.

Knefelkamp, L. L. (1999). Introduction. In W. G. Perry (Ed.), *Forms of ethical and intellectual development in the college years: A scheme*. San Francisco: Jossey-Bass. Koch, C. (1999, February 15). Staying alive. *CIO Magazine*, 38–45.

Kolb, A. Y., & Kolb, D. A. (2005). Learning styles and learning spaces: Enhancing experiential learning in higher education. *Academy of Management Learning and Education, 4*(2), 193–212.

Kolb, D. (1984). *Experiential learning: Experience as the source of learning and development*. Englewood Cliffs, NJ: Prentice-Hall.

Kolb, D. (1999). *The Kolb learning style inventory*. Boston: HayResources Direct.

Kulkki, S., & Kosonen, M. (2001). How tacit knowledge explains organizational renewal and growth: The case at Nokia. In I. Nonaka & D. Teece (Eds.), *Managing industrial knowledge: Creation, transfer and utilization* (pp. 244–269). London: Sage.

Laney, D. (2012). The importance of Big Data: A definition. Gartner. Retrieved 21 June 2012.

Langer, A. M. (2001a). Fixing bad habits: Integrating technology personnel in the workplace using reflective practice. *Reflective Practice, 2*(1), 100–111.

Langer, A. M. (2001b). *Analysis and design of information systems*. New York: Springer.

Langer, A. M. (2002a). *Applied ecommerce: Analysis and engineering of ecommerce systems*. New York, Wiley.

Langer, A. M. (2002b) Reflecting on practice: Using learning journals in higher and continuing education. *Teaching in Higher Education, 7*, 337–351

Langer, A. M. (2003). Forms of workplace literacy using reflection-with action methods: A scheme for inner-city adults. *Reflective Practice, 4*, 317–336.

Langer, A. M. (2004). *IT and organizational learning: Managing change through technology and education*. New York: Routledge.

Langer, A. M. (2005a). Responsive organizational dynamism: Managing technology life cycles using reflective practice. *Current Issues in Technology Management, 9*(2), 1–8.

Langer, A. M. (2005b). *Information technology and organizational learning: Managing behavioral change through technology and education* (1st ed.). Boca Raton, FL: Taylor & Francis.

Langer, A. M. (2007). *Analysis and design of information systems* (3rd ed.). New York: Springer.

Langer, A. M. (2008). *Analysis and design of information systems* (3rd ed.). New York: Springer.

Laudon, K. C., & Laudon, J. P. (1998). *Management information systems: New approaches to organization and technology*. Upper Saddle River: Prentice Hall.

Leavy, B. (1998). The concept of learning in the strategy field. *Management Learning, 29*, 447–466.

Levine, R., Locke, C., Searls, D., & Weinberger, D. (2000). *The cluetrain manifesto*. Cambridge, MA: Perseus Books.

Liebowitz, J., & Khosrowpour, M. (1997). *Cases on information technology management in modern organizations*. New York: Idea Group Publishing.

Lientz, B. P., & Larssen, L. (2004). *Manage IT as a business: How to achieve alignment and add value to the company*. Burlington, MA: Elsevier Butterworth-Heinemann.

Lientz, B. P., & Rea, K. P. (2004). *Breakthrough IT change management: How to get enduring change results*. Burlington, MA: Elsevier Butterworth-Heinemann.

Lipman-Blumen, J. (1996). *The connective edge: Leading in an independent world*. San Francisco: Jossey-Bass.

Lipnack, J., & Stamps, J. (2000). *Virtual teams* (2nd ed.). New York: Wiley.

Lounamaa, P. H., & March, J. G. (1987). Adaptive coordination of a learning team. *Management Science, 33*, 107–123.

Lovallo, D. P., & Mendonca, L. T. (2007). Strategy's strategist: An interview with Richard Rumelt. *The McKinsey Quarterly*, www.mckinseyquarterly.com/Strategys_strategist_An_interview_with_Richard_Rumelt_2039.

Lucas, H. C. (1999). *Information technology and the productivity paradox*. New York: Oxford University Press.

Lucas, H. C. (2005). *Information technology: Strategic decision making for managers*. New York: Wiley.

Mackenzie, K. D. (1994). The science of an organization. Part I: A new model of organizational learning. *Human Systems Management, 13,* 249–258.

MacMillan, I. C. (1978). *Strategy formulation: Political concepts*. New York: West.

March, J. G. (1991). Exploration and exploitation in organizational learning. *Organization Science, 2,* 71–87.

Marchand, D. A. (2000). *Competing with information: A manager's guide to creating business value with information content*. Wiley.

Marshak, R. J. (1998). A discourse on discourse: Redeeming the meaning of talk. In D. Grant, T. Keenoy, & C. Oswick (Eds.), *Discourse and organization* (pp. 65–83). London: Sage.

Marsick, V. J. (1998, October). *Individual strategies for organizational learning*. Lecture presented at the Workplace Learning Institute, Teachers College, New York.

Marsick, V. J., & Watkins, K. E. (1990). *Informal and incidental learning in the workplace*. London: Routledge.

McCarthy, B. (1999). *Learning type measure*. Wauconda, IL: Excel.

McCarthy, E. (1997). *The financial advisor's analytical toolbox*. New York: McGraw-Hill.

McDowell, R., & Simon, W. L. (2004). *In search of business value: Ensuring a return of your technology investment*. New York: SelectBooks Inc.

McGraw, K. (2009). Improving project success rates with better leadership: Project smart. www.projectsmart.co.uk/improving-project-success-rateswith-better-leadership.html.

Mezirow, J. (1990). *Fostering critical reflection in adulthood: A guide to transformative and emancipatory learning*. San Francisco: Jossey-Bass.

Miles, R. E., & Snow, C. C. (1978). *Organizational strategy, structure, and process*. New York: McGraw-Hill.

Milliken, C. (2002). A CRM success story. *Computerworld*. www.computerworld.com/s/article/75730?A_CRM_success-story.

Miner, A. S., & Haunschild, P. R. (1995). Population and learning. In B. Staw & L. L. Cummings (Eds.), *Research in organizational behavior* (pp. 115–166). Greenwich, CT: JAI Press.

Mintzberg, H. (1987). Crafting strategy. *Harvard Business Review, 65*(4), 72.

Moon, J. A. (1999). *Reflection in learning and professional development: Theory and practice*. London: Kogan Page.

Mossman, A., & Stewart, R. (1988). Self-managed learning in organizations. In M. Pedler, J. Burgoyne, & T. Boydell (Eds.), *Applying self-development in organizations* (pp. 38–57). Englewood Cliffs, NJ: Prentice-Hall.

Mumford, A. (1988). Learning to learn and management self-development. In M. Pedler, J. Burgoyne, & T. Boydell (Eds.), *Applying self-development in organizations* (pp. 23–37). Englewood Cliffs, NJ: Prentice-Hall.

Murphy, T. (2002). *Achieving business value from technology: A practical guide for today's executive*. Hoboken, NJ: Wiley.

Nahapiet, J., & Ghoshal, S. (1998). Social capital, intellectual capital, and the organizational advantage. *Academy of Management Review, 23,* 242–266.

Nicolaides, A., & Yorks, L. (2008). An Epistemology of Learning Through. 10, no. 1, 50–61.

Nielsen Norman Group. (2015). User Experience for Mobile Applications and Websites. Retrieved 30 September 2015, from http://www.nngroup.com/reports/mobile-website-and-application-usability/.

Nonaka, I. (1994). A dynamic theory of knowledge creation. *Organization Science, 5*(1), 14–37.

Nonaka, I., & Takeuchi, H. (1995). *The knowledge-creating company: How Japanese companies create the dynamics of innovation.* New York: Oxford University Press.

O'Sullivan, E. (2001). *Transformative learning: Educational vision for the 21st century.* Toronto: Zed Books.

Olson, G. M., & Olson, J. S. (2000). Distance matters. *Human—Computer Interactions, 15*(1), 139–178.

Olve, N., Petri, C., Roy, J., & Roy, S. (2003). *Making scorecards actionable: Balancing strategy and control.* New York: Wiley.

Palmer, I., & Hardy, C. (2000). *Thinking about management: Implications of organizational debates for practice.* London: Sage.

Peddibhotla, N. B., & Subramani, M. R. (2008). Managing knowledge in virtual communities within organizations. In I. Becerra-Fernandez & D. Leidner (Eds.), *Knowledge management: An evolutionary view.* Armonk, NY: Sharp.

Pedler, M., Burgoyne, J., & Boydell, T. (Eds.). (1988). *Applying self-development in organizations.* Englewood Cliffs, NJ: Prentice-Hall.

Penton, H. Material from conversation and presentation at a Saudi business school, 2011.

Peters, T. J., & Waterman, R. H. (1982). *In search of excellence: Lessons from America's best-run companies.* New York: Warner Books.

Pettigrew, A. M. (1973). *The politics of organizational decision-making.* London: Tavistock.

Pettigrew, A. M. (1985). *The awaking giant: Continuity and change in ICI.* Oxford, UK: Basil Blackwell.

Pfeffer, J. (1994). *Managing with power: Politics and influence in organizations.* Boston: Harvard Business School Press.

Pietersen, W. (2002). *Reinventing strategy: Using strategic learning to create and sustain breakthrough performance.* New York: Wiley.

Pietersen, W. (2010). *Strategic learning.* Hoboken, NJ: Wiley.

Poe, V. (1996). *Building a data warehouse for decision support.* Upper Saddle River, NJ: Prentice-Hall.

Porter, M. (1996). What is strategy? *Harvard Business Review, 74*(6), 61–78.

Prange, C. (1999). Organizational learning—Desperately seeking theory. In M. Easterby-Smith, J. Burgoyne, & L. Araujo (Eds.), *Organizational learning and the learning organization* (pp. 23–43). London: Sage.

Prince, G. M. (1970). *The practice of creativity.* New York: Collier Books.

Probst, G., & Büchel, B. (1996). *Organizational learning: The competitive advantage of the future.* London: Prentice-Hall.

Probst, G., Büchel, B., & Raub, S. (1998). Knowledge as a strategic resource. In G. V. Krogh, J. Roos, & D. Kleine (Eds.), *Knowing in firms: Understanding, managing and measuring knowledge* (pp. 240–252). London: Sage.

Rapp, W. V. (2002). *Information technology strategies: How leading firms use IT to gain an advantage.* New York: Oxford University Press.

Remenyi, D., Sherwood-Smith, L., & White, T. (1997). *Achieving maximum value from information systems: A process approach.* New York: Wiley.

Reynolds, G. (2007). *Ethics in information technology* (2nd ed.). New York: Thomson.

Richardson, K. A., & Tait A. (2010). The death of the expert? In A. Tait & K. A. Richardson (Eds.), *Complexity and knowledge management: Understanding the role of knowledge in the management of social networks* (pp. 23–39). Charlotte, NC: Information Age.

Rooke, D., & Torbert, W. R. (2005). The seven transformations of leadership. *Harvard Business Review, 83*(4), 66–77.

Ryan, R., & Raducha-Grace, T. (2010). *The business of IT: How to improve service and lower cost.* Boston, MA: IBM Press.

Sabherwal, R., & Becerra-Fernandez, I. (2005). Integrating specific knowledge: Insights from the Kennedy Space Center. *IEEE Transactions on Engineering Management, 52*(3), 301–315.

Sampler, J. L. (1996). Exploring the relationship between information technology and organizational structure. In M. J. Earl (Ed.), *Information management: The organizational dimension* (pp. 5–22). New York: Oxford University Press.

Sanders, N. R. (2014). *Big data driven supply chain management: A framework for implementing analytics and turning information into intelligence.* Upper-Saddle River, NJ: Pearson Education, Inc.

Schectman, J. (2012, June 7). New EU privacy rules put CIOs in compliance roles. *Wall Street CIO Journal.*

Schein, E. H. (1992). *Organizational culture and leadership* (2nd ed.). San Francisco: Jossey-Bass.

Schein, E. H. (1994). The role of the CEO in the management of change: The case of information technology. In T. J. Allen & M. S. Morton (Eds.), *Information technology and the corporation* (pp. 325–345). New York: Oxford University Press.

Schlossberg, N. R. (1989). Marginality and mattering: Key issues in building community. *New Directions for Student Services, 48,* 5–15.

Schön, D. (1983). *The reflective practitioner: How professionals think in action.* New York: Basic Books.

Senge, P. M. (1990). *The fifth discipline: The art and practice of the learning organization.* New York: Currency Doubleday.

Shaw, P. (2002). *Changing the conversation in organizations: A complexity approach to change.* London: Routledge.

Siebel, T. M. (1999). *Cyber rules: Strategies for excelling at e-business.* New York: Doubleday.

Speser, P. L. (2006). *The art and science of technology transfer.* Hoboken, NJ: Wiley.

Stenzel, J. (2007). *CIO best practices: Enabling strategic value with information.*

Stern, L. W., & Reve, T. (1980). Distribution channels as political economies: A framework for analysis. *Journal of Marketing, 44,* 52–64.

Stolterman, E., & Fors, A. C. (2004). Information technology and the good life. *Information Systems Research: Relevant Theory and Informed Practice, 143,* 687–692.

Storey, J. (1985). Management control as a bridging concept. *Journal of Management Studies, 22,* 269–291.

Swieringa, J., & Wierdsma, A. (1992). *Becoming a learning organization, beyond the learning curve.* New York: Addison-Wesley.

Szulanski, G., & Amin, K. (2000). Disciplined imagination: Strategy making in uncertain environments. In G. S. Day & P. J. Schoemaker (Eds.), *Wharton on managing emerging technologies* (pp. 187–205). New York: Wiley.

Tayntor, C. B. (2006). *Successful packaged software implementation.* New York: Auerbach Publications.

Teece, D. J. (2001). Strategies for managing knowledge assets: The role of firm structure and industrial context. In I. Nonaka & D. Teece (Eds.), *Managing industrial knowledge: Creation, transfer and utilization* (pp. 125–144). London: Sage.

Teece, D.J. (2001). Nonaka, I. & Teece. D. (Eds.) (2001). Strategies for managing knowledge assets: The role of firm structure and industrial context. Sage publications, Thousand Oaks, CA. *Knowledge and Process Management, 10*(4), 125.

Teigland, R. (2000). Communities of practice at an Internet firm: Netovation vs. in-time performance. In E. L. Lesser, M. A. Fontaine, & J. A. Slusher (Eds.), *Knowledge and communities* (pp. 151–178). Woburn, MA: Butterworth-Heinemann.

Tichy, N. M., Tushman, M. L., & Fombrum, C. (1979). Social network analysis for organizations. *Academy of Management Review, 4,* 507–519.

Torbert, B. (2004). *Action inquiry: The secrets of timely and transforming leadership.* San Francisco: Berrett-Koehler.

Tufte, E. R., & Graves-Morris, P. R. (1983). *The visual display of quantitative information* (Vol. 2, No. 9). Cheshire, CT: Graphics Press.

Tushman, M. L., & Anderson, P. (1986). Technological discontinuities and organizational environments. *Administrative Science Quarterly, 31,* 439–465.

Tushman, M. L., & Anderson, P. (1997). *Managing strategic innovation and change.* New York: Oxford University Press.

Van Houten, D. R. (1987). The political economy and technical control of work humanization in Sweden during the 1970s and 1980s. *Work and Occupations, 14,* 483–513.

Vince, R. (2002). Organizing reflection. *Management Learning, 33*(1), 63–78.

Von Stamm, B. (2003). *Managing innovation, design & creativity.* New York: Wiley.

Wallemacq, A., & Sims, D. (1998). The struggle with sense. In D. Grant, T. Keenoy, & C. Oswick (Eds.), *Discourse and organization* (pp. 65–83). London: Sage.

Walsh, J. P. (1995). Managerial and organizational cognition: Notes from a trip down memory lane. *Organizational Science, 6,* 280–321.

Wamsley, G. L., & Zald, M. N. (1976). *The political economy of public organization.* Bloomington: Indiana University Press.

Watkins, K. E., & Marsick, V. J. (1993). *Sculpting the learning organization: Lessons in the art and science of systemic change.* San Francisco: Jossey-Bass.

Watson, T. J. (1995). Rhetoric, discourse and argument in organizational sense making: A reflexive tale. *Organization Studies, 16,* 805–821.

Weill, P., & Ross, J. W. (2004). *IT governance.* Cambridge: Harvard Business School.

Wenger, E. (2000). Communities of practice: The key to knowledge strategy. In E. L. Lesser, M. A. Fontaine, & J. A. Slusher (Eds.), *Knowledge and communities* (pp. 3–20). Woburn, MA: Butterworth-Heinemann.

West, G. W. (1996). Group learning in the workplace. In S. Imel (Ed.), *Learning in groups: Exploring fundamental principles, new uses, and emerging opportunities. New directions for adult and continuing education* (pp. 51–60). San Francisco: Jossey-Bass.

Wideman Comparative Glossary of Common Project Management Terms, v2.1. Copyright R. Max Wideman, May 2001.

Willcocks, L. P., & Lacity, M. C. (1998). *Strategic sourcing of information systems: Perspectives and practices.* New York: Wiley.

Yorks, L. (2004). Toward a political economy model for comparative analysis of the role of strategic human resource development leadership. *Human Resource Development Review, 3,* 189–208.

Yorks, L., & Marsick, V. J. (2000). Organizational learning and transformation. In J. Mezirow (Ed.), *Learning as transformation: Critical perspectives on a theory in progress.* San Francisco: Jossey-Bass.

Yorks, L., & Nicolaides, A. (2012). A conceptual model for developing mindsets for strategic insight under conditions of complexity and high uncertainty. *Human Resource Development Review, 11,* 182–202.

Yorks, L., & Whitsett, D. A. (1989). *Scenarios of change: Advocacy and the diffusion of job redesign in organizations.* New York: Praeger.

Yourdon, E. (1989). *Modern Structured Analysis.* Englewood Cliffs, New Jersey: Prentice Hall.

Yourdon, E. (1998). *Rise and resurrection of the American programmer* (pp. 253–284). Upper Saddle River, NJ: Prentice Hall.

Zald, M. N. (1970a). Political economy: A framework for comparative analysis. In M. N. Zald (Ed.), *Power in organizations* (pp. 221–261). Nashville: Vanderbilt University Press.

Zald, M. N. (1970b). *Organizational change: The political economy of the YMCA.* Chicago: University of Chicago Press.

Index

© Springer Nature Switzerland AG 2020
A. M. Langer, *Analysis and Design of Next-Generation Software Architectures*,
https://doi.org/10.1007/978-3-030-36899-9

Printed in the United States
By Bookmasters